ANGLO-BOER WAR
BLOCKHOUSES

Anglo-Boer War Blockhouses is a fresh analytical look at how the construction of over 9,000 small fortifications during the Anglo-Boer War sought to change its course. The author examines all aspects of the South African blockhouses during the war: how the initial concept of protecting key bridges morphed into mass-produced, low-cost, pre-fabricated forts deployed in long lines across the veld; how they were built, manned and operated in a system designed to defeat roving Boer commandos.

The evolution of the 'blockhouse strategy' used by Lord Kitchener during the guerrilla phase of the war is examined as part of the wider strategy used to bring the war to its conclusion. Detailed analysis through the lens of a military expert finally answers the question 'Did the blockhouses win the war, or were they – in the words of the British Army's nemesis, General Christiaan de Wet – merely the strategy of a blockhead?'

From tracing the use of blockhouses prior to the war, to describing the conditions enjoyed by the average 'Tommy' living and fighting in these structures, to recording their post-war dismantling or preservation, this is a deep dive into a topic previously little explored.

SIMON C GREEN was born in Jersey and educated at Welbeck College. He holds a BSc (Hons) in Technology from The Open University and a Master's Degree in Information Systems from Cranfield University.

Commissioned at the Royal Military Academy Sandhurst in 1978 into the Royal Corps of Signals, he served for nearly 30 years in military appointments in Europe, Brunei, and South Africa. He worked extensively in major military headquarters, giving him a unique insight into the staff who fight wars at a strategic level. He was also involved in training staff officers in operational leadership and war-gaming to develop and hone their war-fighting tactics at lower levels of command. He retired from military service in 2006 and settled in South Africa to pursue his passion for military history and writing.

ANGLO-BOER WAR
BLOCKHOUSES

A MILITARY ENGINEER'S PERSPECTIVE

SIMON C GREEN

To Zodwa
You always support my dreams ... just this one took a while.
Let's both keep dreaming – BIG!

All rights reserved.
No part of this publication may be reproduced or transmitted in any form or by any means without permission from the publisher. The moral right of the author has been asserted.

Copyright © Simon C Green 2020
For reviews please contact the author on simonbsr@gmail.com
and visit his Facebook page @ABWBlockhouses

First published in 2020 by Porcupine Press
ISBN 978-1-9284-5556-1
Second Edition published in 2023 by Blockhouse Books
ISBN 978-0-7961-3254-3

Unless otherwise acknowledged, illustrations are by the author or from the author's collection. Every attempt has been made to trace copyright owners. Any errors made will be corrected in the event of a reprint.

Cover design and page layout: wim@wimrheeder.co.za
Set in 11 point on 15 point, Cambria
Printed and bound by Castle Graphics, Cape Town

Front cover image
Fort Campbell near Brindisi was 10 kilometres south of Fouriesburg, and was one of the more stylish blockhouses, resembling a British castle. The blockhouse, with its unique bartizan or turret, was built by men of the Prince of Wales's Leinster Regiment (Royal Canadians), and completed in September 1901 to protect the Brandwater Basin but was sadly demolished after the war.

Front flap image
Elliot Wood Pattern blockhouse, No. 75, still standing today guarding the Modder River.
Postcard from the author's collection.

Back flap image
A British blockhouse, 'South Africa in 1901', occupied by men of the 21st Battalion,
The Imperial Yeomanry. Photograph courtesy of the National Army Museum, London.

CONTENTS

Preface	10
Introduction	12
What's in a Name?	12
A New Way of Warfare	12
What is a Blockhouse?	15
The Blockhouse System of South Africa	16
CHAPTER 1: A BRIEF HISTORY OF BLOCKHOUSES	18
British Roots in Tudor Times	18
Martello Towers	19
The Period of British Colonialism	20
America (1700–1800s)	20
Canada (1830s)	23
New Zealand Wars (1845–1872)	25
Second Anglo-Afghan War (1878–1880)	27
Suakin Expedition, Egypt (Sudan 1885)	28
Cuban War of Independence (1895–1898)	30
Spanish-American War (1898–1918)	33
CHAPTER 2: A SYNOPSIS OF THE ANGLO-BOER WAR	34
The Historical Context	34
Victoria's Empire	35
Frontier Wars	36
First Anglo-Boer War (1880-1881)	36
The Second Anglo-Boer War (1899–1902)	39
Phase 1: Boer Offensive and Sieges	41
Phase 2: British Relief and Occupation	44
Phase 3: The Guerrilla War	48

CHAPTER 3: BRITISH STRATEGY AND TACTICS — 52

Countering the Train Attacks — 53
    Jack Hindon — 55
Securing the Key Points — 56
Establishing a Constabulary Force — 57
Protecting the Key Towns — 58
Securing the Lines of Communication — 61
Scorched Earth Policy: Removing the Commando Support Structure — 64
Concentration Camps — 67
Denial of Movement — 71
The Drives and New Model Drives — 72
Counter-Mobility Operations — 77
    Converging Mobile Columns — 77
    Night Attacks — 78
    What is Nascent Doctrine? — 79
Armoured Trains — 79
    Percy Girouard — 81
The Birth of Counterinsurgency — 87
Victory in the End — 88
Threat and Counter-Threat & Development of Doctrine — 89

CHAPTER 4: EVOLUTION OF THE BLOCKHOUSE SYSTEM IN SOUTH AFRICA — 92

Forts in the Frontier Wars — 93
Anglo-Zulu War 1879 — 95
First Anglo-Boer War 1881 — 96
Development of the Blockhouse Structures — 98
    Ad Hoc Fortifications and Development — 99
    The Jones' Iron Band Gabion — 99
Masonry Blockhouses — 102
    Elliot Wood Pattern Blockhouse — 102
    Round Blockhouses — 106
    Octagonal Pattern — 107
    Hexagonal Pattern — 110
    Rectangular Pattern — 111
    Square Patterns — 112

Castlettes	113
Hybrids and One-offs	115
Conversions	117
Corrugated-Iron Blockhouses	118
Early Designs	118
Millicent Garrett Fawcett	119
Octagonal Blockhouses	122
Round Blockhouses	123
Blockhouse Development Timeline	123
Cost Effectiveness	128
Evolution into Mass Production	130
Blockhouse Factory Units	132
Expanding Cross-country	134
One Battalion's Blockhouse Building Experience	136
The Whole System Deployed	137
Passes (Poort, Nek and Kloof)	140
SAC Posts	140
Forts and Redoubts	142
Town Guard Forts	143
Summary: How the Blockhouse Lines Spread Across the Country	144
CHAPTER 5: BLOCKHOUSE SYSTEM COMPONENTS	**148**
Human Component	149
Protection System	153
The Blockhouse	153
Fixed Rifle Racks	154
Machine Guns	156
Protection Fences	157
The History of Barbed Wire	157
Connecting Fences	157
Anti-Mobility Ditches	159
Alarm System	160
Alarm Guns	160
Chemical Alarms	161
Electrical Alarms	161
Signal Rockets	161

Ad Hoc Alarms	161
Communication Systems	162
Visual Signalling	162
Physical Dispatches	165
Telegraph and Telephone	165
Wireless Radio in the Boer War	166
Armoured Train Telegraph	168
Wireless Radio	168
Supply System	169
Africans Supporting the War Effort	171
CHAPTER 6: MOBILE BLOCKHOUSES	**175**
Armoured Steam Road Trains	175
Mobile Ox-Driven Blockhouses	180
First World War	189
Second World War	189
CHAPTER 7: LIFE IN THE BLOCKHOUSES	**192**
Boer War Rations	198
Standing Orders	202
Pastoral Care – A Curate's Tale	206
CHAPTER 8: EFFECTIVENESS OF THE BLOCKHOUSE SYSTEM	**211**
Against Bridge Demolition	211
Countering the Train Attacks	213
Protection of the Key Towns	215
Securing the Lines of Communication	216
Denial of Movement	216
Cost Effectiveness	220
The Boer View	222
The British View	224
Recorded Blockhouse Conflicts in the Campaign	225

CHAPTER 9: THE ARCHITECTS OF THE BLOCKHOUSE SYSTEM	**228**
Strategic Architects	228
Lord Roberts	228
Lord Kitchener	233
Lord Baden-Powell	241
Mr Fry of Pretoria	241
The Tactical Designers	244
Major-General Elliot Wood	244
Major Spring Rice RE	251
CHAPTER 10: THE TRANSITION TO PEACE	**256**
Blockhouses Surplus to Requirement	257
Historical Preservation	261
Saving the Blockhouse Heritage	262
The Future of Blockhouses as Monuments	263
Appendix A Blockhouse Lines Constructed during the War	265
Appendix B Calculation of Cost of Blockhouse Materials & Labour	271
Appendix C Rice and Elliot Wood Pattern Blockhouses - Materials Used	273
Glossary	276
Bibliography	286
Endnotes	295
Index	308

PREFACE

Even in 1903 official histories started with 'Why another book about the War?' and since then very many more notable accounts have been added. The answer, for me, is all about fate and circumstances. As a military man for the majority of my working life, battlefields and their stories have always fascinated me, both as part of my military studies and in the lessons they offer on life in general. I have a long-standing interest in how the scars of war are left behind, both on the landscape and in people's individual and personal lives. Secondly I have taken some time to research my genealogical background, tracing my ancestors back through time and their socio-economic stories to around the late 1700s. Never was this more moving than standing by the grave of my Great Uncle Arthur, who lost his life in a German POW camp, and recounting one man's journey across First World War Europe to a group of colleagues on a battlefield tour.

I came to South Africa for the first time in 2002 in the latter years of my military career and completely fell in love with the country and its people, finally settling in Johannesburg in 2007. Having come across a brief mention of the Laingsburg blockhouse in a *Getaway* magazine, this sparked an interest in what these structures were and in what a deep story they might tell. I was a keen geographer and orienteer in my early years and the legacy of the Anglo-Boer War and in particular its footprint of blockhouses seemed a natural topic for my interest and research.

There have been four seminal works of research by Richard Tomlinson, Professor André Wessels, Professor Johan Hattingh, and Professor Wally Peters from which I have been able to gain much knowledge and make a start, but I feel there is a deeper story to tell in terms of how the blockhouse system was deployed and how it fitted into the wider system and operational approach to ending the war. There is also a need to draw a line and make record of the current situation some 120 years after the war – and to add these structures into the battlefield tourism landscape by including the details and access to all the remaining blockhouses.

In my short time in South Africa I have on a few acute occasions been reminded of the deeper legacy of the Anglo-Boer War, and some of its disturbing outcomes. My aim has been to approach this study objectively and with compassion, and to interlace the history with some recent anecdotes and areas that have pulled through into modern life and warfare. Some language of the period will in many cases be read as racist and uncaring, I have tried to be sensitive to this and in some cases removed offensive words or language. Remaining references to ethnic groups are therefore meant to be factual as a historical description and not meant to cause offence in any way.

It is also important to pay homage to those who fought and all those, both military and civilian, who lost their lives, especially so on a continent where at times life seems to be so cheap.

While I love to write, tell stories and unearth the unknown through research and investigation, the engineer in me also wants to relate the quantities of scale involved, and here and there will be some analysis to highlight what effort was made in terms of resources and human time and effort to end a war that was much longer and protracted than was ever anticipated.

INTRODUCTION

WHAT'S IN A NAME?

The war of 1899–1902 is generally referred to as the Boer War or the Anglo-Boer War among English-speaking South Africans, but is also known as the South African War inside South Africa. To the Afrikaner, however, it is the *Anglo-Boereoorlog* or *Tweede Vryheidsoorlog* – the 'Second War of Liberation' (literally the 'Second Freedom War') – or the *Engelse Oorlog* (English War).

It was in fact the second Anglo-Boer War, and was fought from 11 October 1899 until 31 May 1902 between the British Empire and the Dutch-speaking settlers of two independent Boer republics, the South African (or Transvaal) Republic – in the Dutch of the time the Zuid-Afrikaansche Republiek (ZAR), and the Orange Free State (OFS) or Oranje-Vrijstaat. It ended with a British victory and the annexation of both republics by the British Empire; both would eventually be incorporated into the Union of South Africa, a dominion of the British Empire, in 1910.

Occurring some 18 years after the much less well-known First Anglo-Boer War (16 December 1880 to 23 March 1881), in historical circles the war was also termed the last of the 'Gentleman's Wars', referring to the notion of order and chivalry from forgotten times. It occurred on the cusp of a new century when warfare was set to take enormous changes during the Great War of 1914–1918. The world was changing and wars in the latter part of the 19th century were starting to unravel the expansive British Empire, whose soldiers deemed themselves superior in many ways – a stereotypical British trait of arrogance perhaps? The war that linked two centuries in time, tactics and technology certainly heralded an era of innovations on the battlefield, such as X-rays and the wireless radios of Marconi.

A NEW WAY OF WARFARE

During the Cuban War of Independence (1895–1898) fought between the colonial Spanish and the Cubans the concept of 'guerrilla warfare' came into common military terminology. First used (in Spanish *guerrilla* means 'small war') during the Peninsular War in the Napoleonic wars, it referred to the Spanish people's war of liberation against the French using irregular troops and hit-and-run tactics in the mountains of Spain. Over time it came to denote an insurgent, freedom fighter or terrorist, depending on perspective, and of course on which side you were fighting. The next century saw this type of warfare proliferate with the enormous military might of the Americans versus the Viet Cong in Vietnam; the British in Malaya, Kenya and Northern Ireland; and ultimately in the South African fight against the Apartheid government.

The Second Anglo-Boer War typified the nature of warfare when one side is considerably larger and better resourced than the other, which nonetheless has advantages of home soil and a noble cause on which to base a strong fighting spirit and resolution. These classic David and Goliath type struggles would later be termed Counterinsurgency (COIN) Warfare or Asymmetric Warfare, but did not always have the outcome of the Biblical story.

The British Empire had not yet encountered the phenomenon. It had been skilled in and experienced at the grand-scale pitched battles of European campaigns such as Waterloo – on a 'battle-field' with rules of engagement. More recently, with its ever-expanding empire, Britain had become adept at expeditionary wars in which it flexed its colonial might against indigenous or native forces. The British Army of the day came into the Second Anglo-Boer War with a great deal of historical baggage, largely used to winning battles against inferior native oppositions, arrogant, and led by an elitist officer class of upper-class, privileged, and experienced, but elderly campaigners. They had little or no experience of fighting small light forces, adept at living off the land and equipped with excellent fieldcraft skills. The Boer forces would ultimately define the concept of the Commando (Afrikaans: *Kommando*). Additionally the Boers were not an indigenous people but descendants of European settlers, and during the war made a significant difference to the concepts of warfare.

No longer was the British Army fighting stereotypical uneducated or 'barbaric' black natives, as it had considered its opponents in the earlier Cape Frontier Wars or Xhosa Wars (1779–1879) and Anglo-Zulu War (11 January 1879 to 4 July 1879) – this was termed a 'White Man's War'. This aspect engendered a certain respect for each other and an initial understanding that the black population would be kept out of the war, though this proved impossible to resist from either side in the end.

The war was fought by long-term Boer settlers with a passion for the land they had themselves colonised, constantly at odds with the black population; they had a deep-rooted desire and passion to survive. They were pitted yet again against an Empire hungry for resources and with a huge colonial global footprint where failure might mean a cascade of disaster like a 'domino effect' should it be beaten in Africa by a rag-tag bunch of farmers. Ultimately the Empire was not sustainable in the Modern Age; as the political maps were redrawn over and over during the next century, so would the lessons drawn from this conflict shape the strategy, tactics and structures of the British Army in the years following. This conflict might be considered the birth of modern manoeuvres warfare as we know it today. By the end, the concept of this being the last Gentleman's War would be in tatters across the world, as terms such as 'Scorched Earth' and 'Concentration Camp' were also born.

The British Army came to Southern Africa for the second time in 20 years ready to engage with a force it thought it might defeat with relative ease. The story of the war has been well told by many over the years, but this account will focus on the change in emphasis during its three stages, and how the blockhouse system was used as part of a wider strategy to bludgeon the Boers into a spent force and to surrender.

Strategy and tactics were changed and developed 'on-the-fly', and a systemic approach was used to force the Boers into capitulation, as the British moved from defensive into offensive operations. Today the concept of rapid change in the field or during conflict is termed Nascent Doctrine or Doctrine On-the-Fly; for this war the style and nature at the end was very different to that with which it commenced. Over the four years, all bar two infantry regiments in the British Army would serve in South Africa, these were the 15th (The King's) Hussars and the 4th (Royal Irish) Dragoon Guards. Virtually every other infantry regiment would either construct, or serve in, a blockhouse during their tour of duty. So what is a blockhouse and where does it fit into the family of defensive fortifications?

INTRODUCTION

WHAT IS A BLOCKHOUSE?

In researching the many different types of defensive options available and why the blockhouse was chosen for the South African campaign, the author developed a hierarchy of defensive positions. This is not an exact science and for every defined structure there is a hybrid built somewhere. The definition for each level, however, gives a feel for the types of overall options available for static defence of key points, lines of communication and areas of interest. Full definitions are found in the glossary at the back of the book.

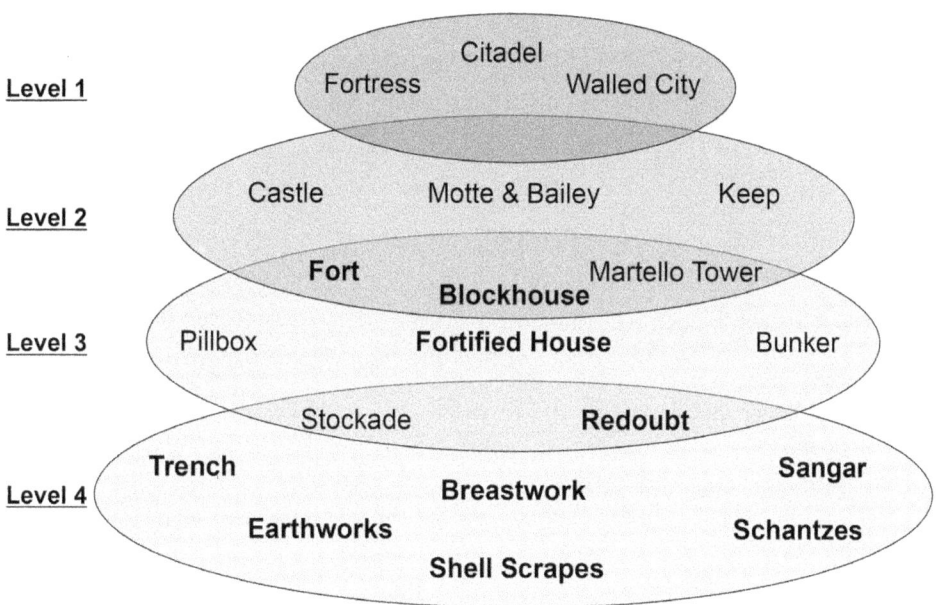

Note: Those in **bold** denote types found during the Anglo-Boer War 1899-1902

A hierarchy of defensive levels of protection.

The types of structures used in South Africa during the Second Anglo-Boer War illustrate that this was more typically a tactical battle, with most of the defensive structures being of a battlefield temporary nature in levels 3 and 4. Those that have best survived this war are contained in level 2 or 3 and are the stone-built blockhouses, which although not considered a permanent structure at the time were constructed to last, and where they have been maintained have survived for 120 years.

Level	Description
Level 1	A permanent type of defensive structure, typified by a Fortress, this would take many years to build by skilled artisans and might be occupied for centuries by different occupants. It would require national strategic sourcing of building material. Nationally strategic in nature, it would afford a high level of protection against a determined siege adversary.
	Examples: Castle of Good Hope might also be termed a Fortress.
Level 2	A permanent type of defensive structure, typified by a Castle, this would take years to build by skilled artisans and might be occupied for several decades by different occupants. It would require national strategic sourcing of building material. Strategic in nature, it would afford a high level of protection against a determined siege adversary.
	Examples: Castle of Good Hope in Cape Town and Zulu War forts
Level 3	A semi-permanent type of defensive structure, typified by a Fort, this would take months to build by semi-skilled artisans or troops and might be occupied for a few years by one set of occupants. It would require sourcing of local building material. Tactical in nature, it would afford a medium level of protection against an adversary.
	Examples: Pretoria Forts; Frontier Forts; Beaufort Martello Tower; remaining stone blockhouses.
Level 4	A temporary type of defensive structure, typified by a trench, this would take minutes or hours to build by troops and might be occupied for a few days by one set of occupants. It would require no building material, relying on what was available on the battlefield. Tactical in nature, it would afford a low level of protection against an adversary.
	Examples: remaining Rice-type blockhouses; remaining battlefield earthworks for British and Boer defences.

Definitions of the defensive levels of protection.

THE BLOCKHOUSE SYSTEM OF SOUTH AFRICA

In tracing the legacy of the Anglo-Boer War blockhouse system and how it formed part of a wider strategy, we shall see the scale of the effort the British Army undertook to build some 9,000 blockhouses across 5,800 miles (9,300km) of South African terrain where at a conservative estimate nearly 60,000 British soldiers and 30,000 African auxiliaries manned their posts for over a year. Based on the author's calculations these may still be on the low estimation of Imperial troops and Africans committed to garrison blockhouse lines. These structures survive in paintings, officers' mess silverware bought by officers who served in them, drawings and diary references, and of course the few surviving physical blockhouses still standing and scattered across the South African landscape.

Of the system of blockhouses completed over 120 years ago, there are approximately 80 of the sturdy stone-built blockhouses remaining as fragments of a network that completely covered what is today the Republic of South Africa. Sadly many of the original 9,000 were reclaimed after the war as building materials, while others have been left to decay in remote areas or those close to towns have been vandalised. There was an effort to preserve some of the masonry blockhouses through the National Monuments Council of South Africa, but a lack of organisational will and funds has meant that further decay has only partially been prevented. The Anglo-Boer War Centenary of 2002 in some small way also assisted in stopping the rot of decay with some restoration and privately funded support being provided in local initiatives. The significance of the Anglo-Boer War and its relics, to a large extent, has been overshadowed by the struggle against Apartheid and there exists apathy by the majority for history and remembrance for this period. Small groups of historians and historical societies work with enthusiasm doing their best to restore these buildings.

The prefabricated and more flimsy corrugated iron blockhouses referred to as Rice Pattern blockhouses, after their inventor Major Spring Rice Royal Engineers (RE), have even fewer surviving examples. Post-war survivors and replicas are recorded and the lines that snaked over the highveld and lowveld recorded in detail. There are map references to 'Blockhouses' and 'Forts' and 'Blockhouse Roads', but investigation often reveals only foundations or piles of spill from the stone-filled cavity walling long since removed. Much of the 120-year-old 'wriggly tin' must still exist in farms and shacks built after the war, scattered across the land in rural areas. Those with a little more foresight built them as replicas on farms or added them as novelty attractions to Bed & Breakfasts, or they are preserved as a museum or tourist attraction. Few remain in their original locations.

This first book aims to capture all the current architectural data and surviving examples through photographic records and searches through the vast array of archives available to deliver the most comprehensive analysis of the blockhouse system to date. The second book, *Anglo-Boer War Blockhouses - A Field Guide*, containing maps and detailed notes should allow much easier access to these battlefield sites and allow a visit in conjunction with other battlefields or journeys throughout South Africa. In addition there are possibilities to conduct blockhouse tours using both books or to add visits to these structures into other activities such as geocaching, hiking and mountain biking.

CHAPTER I

A BRIEF HISTORY OF BLOCKHOUSES

Oh, bird of my soul, fly away now,
For I possess a hundred fortified towers.

RUMI

Blockhouses have a long history as a means of fortification and protection, and are often associated with defences in far-off and remote countries, relics of British and European colonial conflicts. Their roots however lay in the siege and fortification tactics of wars in Europe, where they were extensively used for coastal defence and protection of key sea approaches and harbours. Blockhouses, forts and Martello towers can be visited on travels all over the world, and especially wherever British colonial influence can be found.

BRITISH ROOTS IN TUDOR TIMES

The term 'blockhouse' was most likely first used in Tudor times (1485–1603) and there are some examples remaining in Britain of coastal defences employed against foreign navies and potential invasion. They were defensive structures of various shapes and dimensions specifically built to house roof-mounted artillery pieces that could often traverse 360 degrees. The lower floors housed the magazine and quarters for a small artillery garrison. The earliest known blockhouse, the Cow Tower in Norwich, dates to 1398, but the majority were built in the early part of the 16th century by Henry VIII. Distributed along the east, south and south-west coasts, there are 27 examples which are known to survive in various states of repair, most others are now either destroyed or incorporated into later military constructions.

Early blockhouses were often sited in pairs, and were not built to a standard design, but mainly consisted of a stone tower and a gun platform, which could be semi-circular, rectangular or irregular in shape. English Heritage preserves many of these sites and its website provides a good reference for visitors to sites in Britain.

Blockhouses have been used by the British both at home and in the Age of Exploration during their many colonial exploits. Also, during the 18th century, General Dalrymple employed a blockhouse system in the Caribbean island of St. Vincent (circa 1772–1773) in order to protect British settlers during the British Empire's expansion in this region. Britain was however pre-occupied with costal defence and the threat of invasion by sea, and had built fortifications to counter this threat and that of more modern artillery. The Construction of the Palmerston Forts, to defend key ports and army bases, post-dated the Martello Towers which had become outdated.

MARTELLO TOWERS

Martello towers,[1] sometimes known simply as Martellos, are small defensive fortifications that were built across the British Empire during the 19th century from the time of the French Revolutionary Wars onwards. They stand up to 40 feet (12 metres) high (with two floors) and typically had a garrison of one officer and 15–25 men. Their round structure and thick walls of solid masonry made them resistant to cannon fire, while their height made them an ideal platform for a single heavy artillery piece, mounted on the flat roof and able to traverse, and hence fire over, a 360° circle. A few towers had moats or other batteries and works attached for extra defence.

The Martello towers were used during the first half of the 19th century, but became obsolete with the introduction of the more powerful rifled artillery. Many have survived to the present day, often preserved as historic monuments. In the latter half of the 19th century, there was another spate of tower and fort building during the premiership of Lord Palmerston. These fortifications correctly called the Palmerston Forts, but some confuse them with Martello towers.

The British built three Martello towers in South Africa, one at Simon's Town near Cape Town, one in Cape Town, and the third at Fort Beaufort in the Eastern Cape. The towers at Simon's Town and Cape Town were both built in 1795, the former now situated in the Naval Base and the latter demolished over 100 years ago. The tower at Simon's Town is often claimed as the oldest Martello tower in the world although perhaps it should not be called a Martello tower[2] at all, since it is argued by the purists that it was actually built to provide cover to the already constructed Boetzelaer Battery, and thus has inland-facing embrasures to defend the battery from landward attack, not from the traditional seaward attack of the original design.[3] That said, Rear-Admiral George Elphinstone, who commanded the force that captured the colony and then served briefly as its governor, had served with the Mediterranean fleet off Corsica, home of the earliest Martello tower, in 1794. The British built the tower at Fort Beaufort in 1837, and it is probably the only example of an inland Martello tower ever constructed, another unique feature for South Africa's Martello towers.

THE PERIOD OF BRITISH COLONIALISM

Blockhouses were originally constructed as part of a larger defensive plan, to 'block' access to vital points in the new territory and wherever the settlers had landed. A wide variety of patterns of blockhouses were constructed for defence in frontier areas, particularly South Africa (see Chapter 4), New Zealand, Canada, and the United States.

America (1700–1800s)

The initial period of colonialism in America concerned the arrival of the European settlers and their defence against native American tribes. The French constructed blockhouses for this purpose and to protect their citizens from the British, who were also active in building their own stockades and defensive positions. Constructed from timber or stone (in New England) or of logs banked with earth (in the south and west), they were often part of larger defences of forts or stockades. Palisades frequently surrounded the frontier blockhouses and thus they were technically stockade forts, many of which survive today as well-restored heritage sites, attracting many tourists. Fort Pitt Blockhouse in Pittsburgh, for example, which celebrated its 250th anniversary in 2014, was originally constructed during the French and Indian War. During these early periods of colonisation, there were no artillery pieces available and defence was usually only required from musket fire.

The American Revolutionary War (1775–1783), also termed the American War of Independence – simply the Revolutionary War in the United States – began as a war between Great Britain and the Thirteen Colonies, but gradually grew into a world war between Britain on one side and the newly formed United States, France, Netherlands and Spain on the other. The main result was an American victory and European recognition of the independence of the United States, with mixed results for the other powers.

Central Park Blockhouse No 1, New York.

Further British operations in North America (1812) are reflected in a surviving blockhouse in New York's Central Park, where it is the oldest structure standing. New Yorkers hastily constructed the fort during the War of 1812 in anticipation of a British invasion. Finished in 1814, the fort was part of a series of fortifications in northern Manhattan, which originally also included three fortifications in Harlem Heights (now known as Morningside Heights). It is located on an overlook of Manhattan schist rock, with a clear line of fire into the flat surrounding areas north of Central Park. Officially known as Blockhouse No 1, the fort is the last remaining fortification from these defences. The fort consists of tall stone walls surrounding a small area, inside which a wooden platform would originally have stood. On this platform would have stood a cannon, and the sides house small gun ports. This structure was likely connected to the ground by a small staircase. The British did not attack New York City, and Blockhouse No 1 never saw combat. It was subsequently used for ammunition storage; the top two feet (61cm) of stonework were likely added at a later date as part of this role. The current entrance and staircase were also likely added at a later date, as was the tall flagpole in the centre of the fort.

Blockhouses continued to be constructed in America during the various Indian Wars or Frontier Wars during the 1800s, as America expanded west across the Great Plains towards the Rocky Mountains. Expansion progressed rapidly with the development of the railroad and the discovery of gold. The Army built a network of forts as expansion spread west, comprising large forts, stockades and temporary cantonments or camps. Although the Army was relatively small at 25,000 men[4] during this period, the distribution of fortifications across the vast plains was extensive, if spread a little too thinly to be effective. Forts were often large, and required barracks and stabling. They had wooden log stockade walls with inner buildings, which often had one or two blockhouses incorporated for their inner sanctums or at opposite corners of the square fort. By 1860 there were 73 forts and cantonments on the frontier, and as the war drew in more Indian tribes this expanded to 116 posts by 1867.[5]

Stockade walls were recommended to be at least 12 inches (30cm) thick and supported a walkway from which to gain the best firing position and for observation across the plain. Earthworks often supported the walls. The blockhouses were usually two stories, often with the second floor overhanging the base or offset by 90 degrees in order to provide better all-round defence and enfilade fire along the walls. They often incorporated small cannon, such as a 12-pounder howitzer, and had larger louvered apertures from which to fire, usually on the first floor, (US second Floor). Fort Fairfield Blockhouse[6] in North Maine, USA, built with fat, square, rough-hewn timbers, sits right in the middle of town on a busy street, but 200 years ago, this was a frontier spot, and at the epicentre of a tense conflict known as the Aroostook War.

Blockhouses like Fort Fairfield were well designed to live in, often having an internal stove and chimney for cooking warmth. The style of this blockhouse was sufficient for defence against the Native American tribes armed with muskets and the bow and arrow, but something better was required for the American Civil War, where each side had large cannon. In these designs the walls were built to a double thickness. One such blockhouse is the Crockett Blockhouse (1855–1856), which was built in Washington State on the American west cost during the Yakima Indian War. Re-sited from its original position and restored in 1938, it was originally one of a pair enclosing Colonel Walter Crockett's farm, the other blockhouse having been sold and moved to Seattle in 1908.

A BRIEF HISTORY OF BLOCKHOUSES

*Fort Fairfield Blockhouse in North Maine.
Courtesy of Nick Woodward / MPBN.*

Canada (1830s)

Blockhouses were built by the British in the 1830s on the Bois Blanc Boblo Island at the mouth of the Detroit River during the Canadian rebellion called the Patriotic War. Fears of civil insurrection in Upper Canada lasted for some time after the rebellion of 1837 had been put down. Evidence of the military authorities' continuing concern for maintaining order is found in a trio of blockhouses built in 1838–1839. They were intended to guard the roads to Toronto from the north, rather than for defence against hostile Indians who might be approaching Toronto from the Don Valley as was asserted in 1929 by HH Langton, the biographer of Sir Daniel Wilson. The Canadian authorities have realised the potential tourism value of these blockhouses and have been active to restore and retain 20 fine examples for posterity, two of which are shown here.

*The Lockmaster's House Blockhouse, Rideau Canal, Ontario.
Courtesy of D Gordon & E Robertson.[7]*

The Lockmaster's House Blockhouse at Rideau Narrows lock was built in 1832, on the Rideau Canal, Ontario. The blockhouse never had to serve its military purpose. It is possible that during the rebellion crisis of 1837–1838 members of the second regiment of the Leeds Militia may have stayed in the blockhouse. By 1856 the blockhouse was serving exclusively as the lockmaster's house. In 1888 it underwent extensive renovations. Large framed additions were added, giving it the appearance of a farmhouse and all but obliterating any visible evidence of the blockhouse. In 1967 and 1968 the blockhouse underwent extensive renovations back to its original structure, which has a stone-built ground floor and a timber first floor.

The St Andrews Blockhouse overlooks the harbour and waterfront of St Andrews, New Brunswick. The blockhouse is associated with the coastal defence of British North America against American privateers and military forces during the War of 1812, but never saw action. Twelve similar structures were built, but only the St Andrews blockhouse still stands. It was repaired in the 1990s following a fire. It is a good example of an operational 19th century military design. Its defensive details include numerous small loopholes for musketry and portholes for small-calibre artillery, and an overhanging first floor.

*The St Andrews Blockhouse, New Brunswick.
Photo courtesy of John Stanton.*[8]

New Zealand Wars (1845–1872)

At least 32 redoubts, stockades and blockhouses were built in New Zealand from 1840 to 1848, during the first eight years of British Crown rule. The structures, built from stone and wood, were for protection for the settlers during the war that was largely fought between tribes over rights to tribal lands. The introduction of the musket – the conflict was also referred to as the Musket Wars – changed these inter-tribal battles and had a decimating effect on some of the tribes. Today only seven sites remain, with only two having visible ruins.[9] The colonial government and the settlers constructed them after the signing of the Treaty of Waitangi in 1840, but many were abandoned or destroyed after falling into neglect. In addition to defending the local population, the structures were designed to protect a strategic site or to defend a line of advance. They were built to house muskets for defence and similarly only had to protect the inhabitants from musket fire, the effective range of which during this period was approximately 98 yards (90 metres).[10]

The New Zealand Wars, sometimes called the Land Wars or the 'Māori Wars', were a series of armed conflicts that took place between 1845 and 1872. The latter name originated from the 19th century British practice of naming their wars after their foe, such as the Zulu Wars, or the Indian Mutiny. The wars were fought over a number of issues, the most prominent concerning Māori land being sold to the settler population, and was fought between European settlers and Māori.

Wallaceville Blockhouse, New Zealand.[11]

Only those blockhouses constructed at a later period have remained and are usually well preserved. An example is the Upper Hutt Blockhouse, or Wallaceville Blockhouse, located in Upper Hutt. The American-style fort was built during the Maori Wars, and is one of very few of its type remaining in New Zealand. It was erected in 1860 in response to a fear held by local settlers that the conflict between Māori and the Crown over the disputed sale of land at Waitara, Taranaki, would escalate.

The wooden blockhouse has a double-skin timber-clad frame with shingle to provide protection against rifle fire. Loopholes were built into the lower story for defenders to return fire. The blockhouse was surrounded by a much larger defensive earthwork, which was later flattened to form part of the school playing fields now surrounding the blockhouse. A small militia occupied the blockhouse; however, the attacks never happened and they left the blockhouse in 1861. Between 1867 and 1868 it was used as a police house, then a courthouse and residence until 1880, and was declared a historic reserve in 1916. In 1927 the building was substantially repaired and windows were added on the inner side of the L-shaped structure. In 1980 the blockhouse and the accompanying land was classified as a historic reserve under the Reserves Act 1977. Soon after that the New Zealand Historic Places Trust was appointed to control and manage the blockhouse. From 1953 until the late 1990s local boy scouts and girl guides used the building,[12] and today it serves as the headquarters for a service club, illustrating how historic buildings, even those constructed from wood, might be preserved for modern day use.

Second Anglo-Afghan War (1878–1880)

In the 19th century the British were engaged in military conflicts with Afghanistan due to fear of Russian encroachment on their Indian colony and internal divisions within Afghanistan. During the Second Anglo-Afghan War British forces commanded by General Frederick Roberts occupied the area around Kabul and endeavoured to gain control of the imposing hills around the city. An extensive programme of defensive works was begun which included blockhouses linked together with fortified walls and trenches along the Bimaru Heights, and new roads for the transportation of artillery. Roberts's experience securing Kabul and using blockhouses most likely gave him the imagination to deploy their initial use in the Anglo-Boer War.

The blockhouse on the western end of the Siah Sung ridge was part of a larger fortification aptly named Fort Roberts, and capable of housing 1,000 men.[13] Photographs of the time show the stone structures to be very similar to some of those built around Johannesburg, with multi-storeyed, loopholes and a first-floor entrance. The blockhouse also has a unique set of brattice or hooded firing-points for firing down onto the lower wall, resembling large air vents. These hooded firing-points seem typical of a particular design used in Afghan fortifications[14] in and around Kabul from the 16th century onwards, and in forts such as the Red Fort in Delhi, India, at the same time. They were constructed of stone, burnt-brick and plaster and were relatively light in construction. None survive today on any of the remaining Afghan ruins.

Blockhouse at Siah Sung, 1879. Courtesy of the British Library.[15]

Hooded firing point in use to fire vertically on attackers.[16]

Many of these forts were still in existence when the British forces were there, in Roberts's time, and would have influenced British blockhouse design in future campaigns. The need for some sort of enfilade fire to cover the walls from close quarter attack exists for every fixed fortification, so it will similarly be shown for blockhouses built in South Africa. The designers of blockhouses took their inspiration from a variety of sources, their Royal Engineer training, British medieval castles and fortifications seen in foreign lands such as Afghanistan.

In addition Lieutenant EH Bethell RE (later to be Inspector of Blockhouses in South Africa) served as Officer Commanding 5th Company Bombay Sappers and Miners and would have constructed many of the blockhouses used in the Afghan theatre of operations.

Second Suakin Expedition, Egypt (Sudan 1885)

Suakin was an important conflict from a British expeditionary warfare perspective as many of the British officers who were to shape the use of blockhouses in South Africa in 15 years' time were also present as commanders or engineers in this conflict. This campaign refers to two expeditions; the first being in February 1884, through to the second expedition in March 1885, as part of the Anglo-Egyptian War of 1882–1885 where Lord Kitchener and Lieutenant Colonel Elliot Wood were both Royal Engineer officers. The experience they gained in this conflict would have a significant impact on their thinking and use of blockhouses in the South African campaign, although the enemy and terrain were very different.

As a Brevet Lieutenant Colonel, Elliot Wood led the 17th (Field) Company RE in the Suakin Expedition and was responsible for a railway line from Suakin to Berber (although it was never constructed).[17] It was during his service at Suakin that he developed a notable blockhouse, Fort Handub. These blockhouses, constructed from local stone, sandbags and wood, were often built to command key ground and had tall towers from which to provide effective manual signalling. The buttressed and battered walls are reminiscent of medieval castles (a key indication of where the designers and engineers drew their inspiration).

Fort Handub, Suakin, Egypt.
Reproduced by permission of Durham University Library and Collections.[18]

CUBAN WAR OF INDEPENDENCE (1895-1898)

The Cuban War of Independence was the last of three liberation wars that Cuba fought against Spain on the island. The Spanish General Arsenio Martínez Campos tried the same strategy he had employed in the Ten-Year War (1868-1878), constructing a broad belt of defences across the island, called the 'trocha', about 50 miles long (80 kilometres) and 220 yards wide (200 metres).

Once the war degenerated into a war of sabotage, banditry and guerrilla tactics the Spanish built the trocha to cut off the prosperous western part of the island and concentrated the rural population in fortified towns. It tore directly across the landscape through dense jungle and made no allowances for topography or the road system. The cut logs and torn-up tree roots were piled on either edge of the clearing to form a barrier higher than a man. In between the clearing there was a military road with barbed wire entanglements connecting a line of fortifications or blockhouses. There were three different types built: large fortifications, blockhouses and small fortifications.

The large forts, placed at half-mile (800 metre) intervals, were two-storey buildings comprising a cellar below and a watchtower above, made of stone and painted a glaring bright white. On a good clear day a man in the watchtower might see three forts down the line. In between these forts were two-storey blockhouses built of a mud lower part and an overhanging upper story of wood. Finally between each blockhouse and each fort were built tiny forts of mud and planks surrounded by a ditch, which contained five fighting men. They were similarly encased in stakes and barbed wire for self-protection and to prevent insurgents passing through. Where the jungle was less dense, bombs were placed with wires connected to the forts and blockhouses for remote detonation if anyone passed by.

*Spanish blockhouse on the outskirts of Santiago, Cuba,
the style of which is very similar to those built in South Africa.*

The trocha was examined at the time by Richard Harding Davis (1864–1916), the best-selling American journalist and author who declared that: 'every sheet of armour plate, every corrugated zinc roof, every roll of barbed wire, every plank, beam, rafter and girder, even the nails that hold the planks together, the forts themselves, were shipped in sections, which are numbered in readiness for setting up the ties for the military road which clings to the trocha from one sea to the other...'[19] In a final summary Davis draws an analogy to another British conflict that was to provide background interest to the blockhousing in South Africa: 'A trocha in an open plain, as were the English trochas in the desert around Suakin, makes an admirable defence, when a few men are forced to withstand the assault of a great many, but fighting behind a trocha in a jungle is like fighting in an ambush, and if the trocha at Moron is ever attacked in force it will prove to be a Valley of Death to the Spanish troops.'[20] In effect the men inside the defensive lines felt more like they were in an ambush themselves as the jungle and the piled trees provided ample cover from which to attack the defenders. The Trocha was thus not at all effective: as Davis eloquently described: 'insurgents apparently crossed at will with the ease of little girls leaping over a flying skipping rope'.[21] This was largely achieved by intelligence provided to the insurgents by local peasants living close to the trocha where crossing was possible.

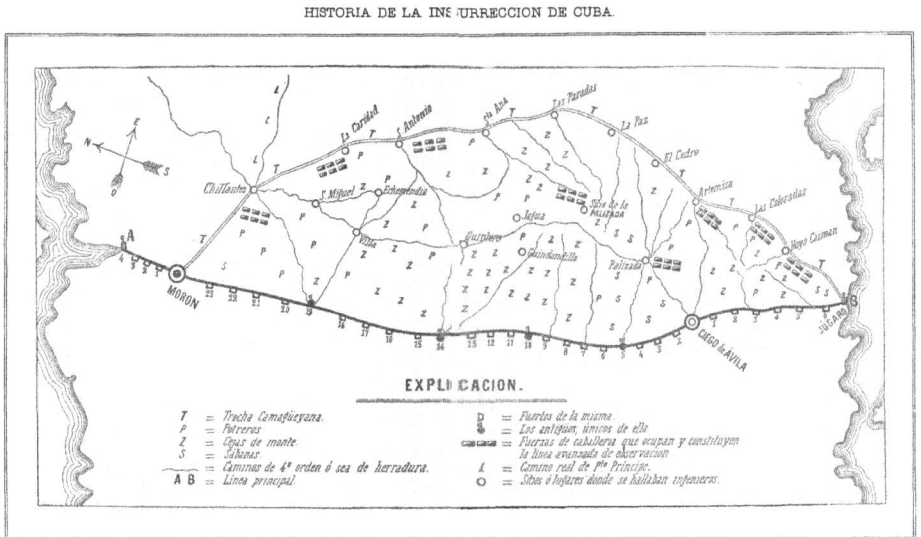

Map of the blockhouse line in Cuba from the Spanish Wars.[22]

Blockhouses similar to the one here were established by the Spaniards at all important points on the island, and at definite intervals along the trochas. They were frequently surrounded, at a convenient rifle range, with a network of barbed wire, rendering them exceedingly difficult to approach. To supplement the trocha there were other blockhouses constructed all across the island. These were deployed in straight lines, zigzag lines or circles, but generally dominating the land from high ground and hilltops. They could be 400 yards (365 metres) apart or within a stone's throw of each other, depending on the topography of the land, unlike the trocha, which merely cut a swath across the island. Small bullet-proof fortresses were used to house troops and to act as a daytime patrol base, with troops returning to these enclaves at night.

In addition to the reinforced trocha; 'concentration camps' were also used to deprive the rebels of civilian support with the aim to eventually grind them down, and armoured trains were also used to protect and control strategic railway lines. This strategy of reinforced trochas, blockhouses, armoured trains and concentration camps would however ultimately fail, unlike the same strategy employed by Kitchener in South Africa. It was a policy called 'Reconcentration' whereby rather than trying to pull the rural populations into the cities, the Spanish turned the blockhouse system inside out and went out into the countryside to concentrate the rural dispersed population with their livestock at specified points. New trocha lines were then constructed and used to beat the rebels against them, much as the British would do in South Africa some five years later.

The Spanish, numbering some 360,000, while holding possession of these fortifications, failed in the main part to disrupt or capture the 50,000 Cuban insurgents, who continued to exert their influence over the civilian population. Other factors also contrived against the Spanish, such as disease brought about by concentrating the army with the population – and then the entering of the United States guaranteed an end to the war.

Of interest is that who should also make an unlikely appearance in this war, but the young Winston Churchill. Officially he was on leave of absence from his regiment at the time and went there in 1895 at the age of 21, gathering information as a journalist-cum-'military observer'. Such was the similarity of the tactics and methods used in Cuba that there is potential that Churchill's articles influenced their future use in the Anglo-Boer War a few years later.

SPANISH-AMERICAN WAR (1898)

The Spanish-American War was a conflict in 1898 between Spain and the United States, effectively the result of American intervention in the on-going Cuban War of Independence. American attacks on Spain's Pacific possessions led to involvement in the Philippine Revolution and ultimately to the Philippino-American War.

The year 1898 saw the 24th Infantry Regiment (United States) deployed to Cuba as part of the US Expeditionary Force in the Spanish-American War. Elements of the 24th participated in the storming of the Spanish fortress in the Battle of El Caney, Cuba. At the climactic battle of San Juan Hill, supported by intensive fire from the Gatling gun detachment, units of the 24th Infantry, accompanied by elements of the 6th and 13th Infantry Regiments, assaulted and seized the Spanish-held blockhouse and trench system atop San Juan Hill. This American unit has the unique position of their unit insignia containing an image of the blockhouse. Today the site can be visited, where there is a reconstructed blockhouse flanked by Spanish and American artillery pieces.

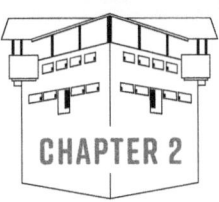

CHAPTER 2

A SYNOPSIS OF THE ANGLO-BOER WAR

*The Boer War was other than most wars.
It was a vast tragedy in the life of a people,
whose human interest far surpassed its military value.*

**GENERAL JAN SMUTS,
PREFACE TO *COMMANDO* BY DENEYS REITZ**

It is not the author's aim to retell the entire story of the war, which has been achieved in countless great works, but merely to set the strategic and cultural backdrop, and outline the plan against which the Anglo-Boer War plays out. It is important to understand the key strategic position of the British Empire of the day and how this channelled the generals into their various strategies and tactics, and ultimately why the blockhouse system as part of a greater strategy was conceived and considered necessary.

THE HISTORICAL CONTEXT

The Anglo-Boer War was the military climax of a long-standing rivalry between the British and the Afrikaner, which had started centuries before in the Cape Colony. Although there was a union after the war Boer and British rivalry simmers to this day for some. The Kaap de Goede Hoop was originally settled in 1652 by the Dutch East India Company for trading purposes, as a resupply point between Jakarta in the Dutch East Indies and the Netherlands, and the farming communities it encouraged were the forefathers of the Boers, or Afrikaners, a white ethnic group in the southern African continent.

The Cape was first invaded by the British in 1795, prompted by the need to protect this strategic port and trading route from the French during the Napoleonic Wars. After briefly reverting to Dutch rule it became a full and integral part of the British Empire in 1814, when the Dutch government formally ceded sovereignty to the British under the terms of the Convention of London.

The Cape Colony continued to evolve and expand largely under Dutch influence but with a British garrison in place to ensure its security. The local Khoikhoi cattle herders had been subjugated over time and had become the chief source of colonial labour, together with slaves imported from the East and via the African slave trade. Whilst the British continued to colonise and expand out to the eastern border, they also began to introduce rudimentary rights for the Cape's black population and on 1 August 1834 officially abolished slavery across most of the British Empire. One of the exceptions was the Cape Colony where it was delayed for four months until 1 December. The Act apprenticed slaves to their masters for a period of four years. This enabled them to learn trades and afforded a transition period for the owners.[23]

The Boers relied on their slaves for labour and this, combined with resentment of the imposition of English language and culture, caused them to seek independence. The mass emigration from 1834 to 1846 was known as the Great Trek: trains of ox-wagons trekked northwards inland where eventually the two Boer Republics of the Transvaal and Orange Free State were established. By 1854 Great Britain formally recognised the independence of these two Republics.

Victoria's Empire

It is worth setting this period in a global context as in 1837 Great Britain entered the Victorian Era,[24] a period of enormous strategic competition and growth, which built Britain's Empire. As the world's first industrialised nation, Great Britain and her empire of colonies were expanding on a path to counter threats from other European nations and to become a dominant factor in global commercialisation. There was also a drive to colonise and civilise native populations, which all contributed in the race to acquire vast tracts of Africa for its resources and strategic ports on the vital global trade routes. The Victorian Era in this regard was also dominated by military campaigns; there was not one year in the entire 63 years of Queen Victoria's reign when the British Army was not fighting somewhere around the globe. In fact 72 different military campaigns were fought largely as small expeditionary wars against native armies, with only one or two notable exceptions such as the Crimean War against Russia. Once established, these self-governing colonies relied heavily on British sea power and small land force detachments of British officers and men with locally recruited soldiers to ensure their security. Thus the Army was large, spread out globally with a strategic focus on India, and geared for expeditionary colonial campaigns.

Frontier Wars

Also in the colonial mix were the continued and on-going battles with the indigenous populations of South Africa. The expansion of the Voortrekkers on the Great Trek, with battles against the Ndebele and Zulu nations along the way, was a bloody and epic nation-building success culminating in the Battle of Blood River in 1838 and the founding of the Natalia Republic, which lasted until 1843. The British, having spent the best part of 100 years attempting to subdue the Xhosa of the Cape's Eastern Frontier, were also expanding into Zululand. There they came head-to-head with King Cetshwayo's Zulus, desperate to hold on to the last vestiges of their power during the Anglo-Zulu War (1879), which with its epic encounters at Isandlwana and Rorke's Drift led to the eventual destruction of the Zulu kingdom. The African chiefdoms were clearly in the middle of the fractious Boer and British nations and in all cases the Boers considered themselves unfairly treated.

Tensions between the British and the Boers continued to simmer as the 19th century progressed, fuelled by three factors.[25] Firstly, there was the need to secure the trade routes around the South African Cape to India, which lay at the strategic heart of the Empire. Then there was the discovery of diamonds at Hopetown, near the Orange Free State border (1867). Finally, there was the race to counter threats from the other European colonial powers, namely Germany, France, Belgium and Portugal. European relationships were tense as the race to colonise Africa continued.

First Anglo-Boer War (1880–1881)

Britain had annexed the Transvaal in 1877 as part of a strategic plan to bring together under a confederation all the southern African states. This was unopposed by the Boers; although not universally popular, it was allowed due to drought, a lack of funds to raise an army and threats from neighbouring Zululand. However, the defeat encountered at Isandlwana revealed to the Boers – now thirsting for the reinstatement of their independence – that warriors armed only with primitive weapons could inflict a humiliating defeat on the mighty British Army.

The spark in the gunpowder came when a Boer named Piet Bezuidenhout refused to pay an illegally inflated tax. Government officials seized his wagon and attempted to auction it off to pay the tax on 11 November 1880, but 100 armed Boers disrupted the auction, assaulted the presiding sheriff, and reclaimed the wagon. The first shots of the war were fired when this group fought back against government troops who were sent to capture them.[26]

On 13 December a great meeting of burghers pledged to restore the republic. On 16 December independence was declared at Heidelberg and on the 20th a British column called back to Pretoria because of rising tensions was 'ambushed' at Bronkhorstspruit (although the ensuing battle might be better classed as a meeting engagement). The British commander of the force was warned of an impending Boer attack, but they met the Boers under a white flag and were warned not to proceed, else any move may be considered an act of war. The British took umbrage at being told what to do, and were ill-prepared for the Boer attack. The initial engagement was a devastating defeat for a poorly prepared column by a capable and determined Boer force. The British carried insufficient ammunition, failed to conduct adequate scouting, were 'at the halt' and did not deploy into any form of defensive position. Once action commenced they lost their officers early in the engagement and failed to adequately adjust rifle sights to the correct range for the Boer positions.[27] Their uniforms still consisted of red tunics and white helmets and presented the Boer marksmen with very distinctive targets, at which they were able to engage at considerable ranges, up to 1,000 yards (900 metres). In an engagement that lasted only about 15 minutes the casualties were very high: the British had 77 killed in action from a total of 157 casualties, while the Boers lost 2 killed in action and 5 casualties.[28]

The war, albeit brief (it only lasted 10 weeks), had two other significant actions, which apart from dictating the outcome of this first Anglo-Boer War also shaped the future for the next. The scars that were to remain in British hearts were to severely dent morale and pride. Several junior officers who were to fight in these battles became the generals and staff officers for the next Anglo-Boer war.

At Laing's Nek on 28 January 1881, Major-General Sir George Pomeroy Colley had reluctantly rushed to send reinforcements forward to relieve the besieged garrisons in the Transvaal. The Natal Field Force under Colley's command attempted to break through the Boer positions on the Drakensberg mountain range at Laing's Nek to relieve these garrisons. The British were repelled with heavy losses by the Boers under the command of Piet Joubert, who sited his forces in well-entrenched positions on both sides of the road. Of the 480 British troops who made the attack, there were 190 casualties with 144 dead. Furthermore, sharpshooting Boers had killed or wounded many senior officers, including all but one of Colley's staff. The Boers had lost only 24 casualties, with 14 dead. Two Victoria Crosses were awarded in this action for acts of outstanding bravery, but it was another blow to the British.

The British rout however was not over, and although peace negotiations were already in progress, Colley, ever keen to avenge these humiliating defeats and restore his reputation, decided to take the steep-sided but flat-topped Majuba Hill, on 27 February 1881. The hill was reported vacant during the night, having only a daylight hours picket in place. The position overlooked Laing's Nek and its occupation, he surmised, would result in a Boer retreat. He selected an unconventional mix of 600 men comprising four different regiments of both short-service recruits and long-service veterans, and a section of the Naval Brigade. This mix later led to problems of unfamiliarity with senior officers, reduced cohesion and a lack of esprit de corps. On reaching the summit at daybreak after an eight-hour climb, the troops were alleged to be exhausted and failed to sufficiently develop a defensive position. Later in the morning they were routed by a Boer force able to use the ground to their advantage and engage the British from rocky positions. Over 270 British troops were casualties, including General Colley, who died, while the Boers only had two men killed and five wounded. The defeat left a deep scar on the British who would in years to come commemorate the defeat with the battle cry 'Remember Majuba!'. The British had yet again underestimated the Boer fighting ability, skill and spirit, and were beaten for the third time.

The British government, under Prime Minister William Gladstone, realised that any further action would require substantial troop reinforcements, and it was likely that the war would be costly, messy and protracted. Unwilling to get bogged down in a distant war, the British government ordered a truce, and the Transvaal became a Boer Republic again.

Key lessons from this short and sharp introduction to fighting the Boers, and which should have been learned when they next encountered them were:
- The British Army was not at all happy with the politicians' decision to quickly sue for a peace with the Boers. Lord Roberts was on his way to South Africa with reinforcements, but by the time they arrived the war was over. The military discontent and dented pride however masked the effect this victory had on the Boers. They considered that British had little

appetite for war in the region, and British generals failed to take heed of this resolute belief in the first engagements of the next war.
- The Boers took away from these battles the firm belief that they could defeat the mighty British Army. The commandos were on ground they knew and used it to good advantage, and in well-prepared positions were able to use their keen marksmanship skills to engage obvious red tunic targets at over 1,000 yards (900 metres).
- The British were still tied into rigid tactical drills and configured to think, fight and command as if they were fighting a 'native force'. They were not used to engaging a similarly equipped enemy, and one who could use its weapons more effectively. Although 'only farmers', the Boers outfought the British, who still relied heavily on their officers for range instruction and orders to fire in volleys. In short, the British Army was ill-prepared for modern warfare.

THE SECOND ANGLO-BOER WAR (1899-1902)

Eighteen years later the prelude to the Second Anglo-Boer War was set against a background of strong Afrikaner nationalism, for an Afrikaans nation from Table Bay to the Zambesi, and British imperialism as part of their continued expansions in Africa. The clash between these two very different nations, peoples and cultures had been previously exposed during the First Anglo-Boer War, but the matter had been quickly resolved. It had however left some misconceptions regarding the resolve to fight over African soil on each side that would again be tested in this much longer war. British interests had not been decisively concluded during the First Anglo-Boer War; the politicians had fudged the conclusion, leaving the Boers to believe that the British had no stomach for fighting over African soil. The British conversely had been resoundingly beaten at Majuba and had the British Army had old scores to settle, once battle commenced, but were arrogant in their view of Boer capabilities, and unprepared for an extended campaign.

The Afrikaners had for years been trying to escape British rule by trekking north out of the Cape region and now finally they had their own republics. The discovery of diamonds in Kimberley in 1870 and gold on the Witwatersrand in 1886 had however started to destabilise the already tense relationships between Britain and the independent Boer Republics.

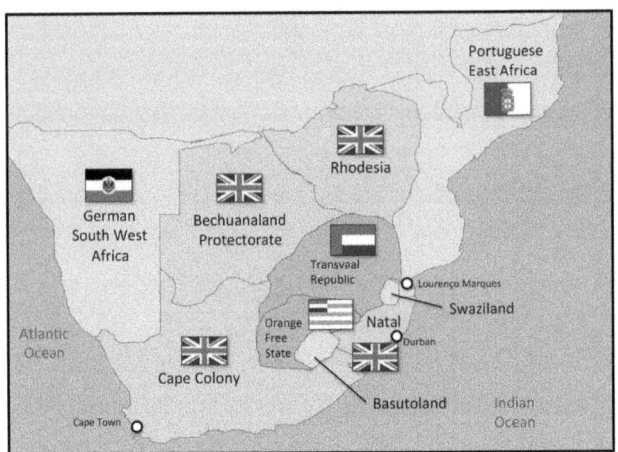

Political map of Southern Africa in 1899.[29]

There was a huge influx of labour from the African communities into the new developing city of Johannesburg and also of skilled expatriate mining labour from Europe. Johannesburg, or eGoli (the 'city of gold') as the locals called it, became a powerhouse of economic development, presided over by the Randlords, rich and powerful mining magnates, such as the imperialist Rhodes and the financier Alfred Beit.

These outsiders or *'uitlanders'* came into the Republics with few rights, and they were denied the right to vote under the Afrikaner administration, despite having become the majority in Johannesburg. In addition they were paying taxes, doing most of the hard work down the mines but denied management posts or access to services and schools. They were growing progressively more restless and vocal in their opposition. Many of them were British and the British government was watching and waiting.

The overtly colonial Rhodes had unofficially consented to the Jameson Raid on 29 December 1895, with the aim of supporting the uitlanders and overthrowing the ZAR. The raid ultimately failed; Rhodes was forced to resign as Prime Minister of the Cape Colony and tensions were further escalated. To others it seemed the British had violated their obligations under the London Convention of 1884, not to interfere in the domestic business of the Transvaal. The failed raid had also bolstered support for Kruger as President of the ZAR, and rallied support for an independent Orange Free State.

The years between the Boer Wars had seen the Boer republics considerably rearm with the latest technology, much at odds with the British belief that this would be an easy adversary to defeat. Tensions rose further when Sir Alfred

Milner, the Governor of the Cape, rejected Kruger's limited offer of reform at the Bloemfontein Conference in June 1899. He was in any case deeply distrustful of the Boers and considered a quick and easy war a better way to settle matters than any agreement of appeasement to prevent war.

The British government, reacting to the growing tension, sent 7,400 troops by sea, of course, to Natal to bolster their defences, which prompted the Boers to issue an ultimatum for them to be withdrawn.

Phase 1: Boer Offensive and Sieges

Boer forces fielded a total fighting strength of 38,000 burgher or citizen soldiers formed into localised units known as commandos, when they attacked. Permanent forces of the two Republics only included small police and artillery units. They had however been preparing for conflict and had gone through a process of militarisation, buying modern weaponry in the form of Creusot and Krupp artillery pieces and 40,000 German Mauser rifles. Those British officers who had served at the end of the First Anglo-Boer War at least knew of an intent to fight the British again. Sir Frederick Darley testified at the Royal Commission of Inquiry in 1902[30] and remarked that 'even in 1881, when I left the country the Boers had stated they would attack when it considered England was weak'. This had been the idea that had pervaded the minds of leading Boers in their pursuit of certain rights and privileges. Further reports in 1896 had also given details of the Boer militarisation and modernisation programme, yet the British Army seemed ill-prepared for the early aggression.

At the commencement of the Boer offensive the State Artillery of the South African Republic, which included the Fortress Artillery Corps, was manning the four Pretoria Forts and the Johannesburg Fort. The Corps was commanded by a lieutenant colonel and consisted of eight Krupp 75mm field guns, six 75mm Creusot field guns, twenty 37mm Maxim-Nordenfeldt Pom-Poms, four 120mm Krupp Howitzers, and the famous four 155mm Creusot Fortress guns (or Long Toms).[31] The unit was divided into four divisions and sent to Natal, Mafeking, Colesberg and Kimberley. The Orange Free State Artillery Corps had 5 officers and 159 other ranks with a reserve of some 400 more. Its order of battle consisted of twenty-four assorted artillery pieces – fourteen 75mm Krupp breech loaders, five 9-pounder Armstrong rifled muzzle loaders, one 6-pounder Whitworth rifled muzzle loader, one 37mm Krupp single-loader, three 3-pounder Whitworth muzzle-loading mountain guns and three Martini-Henry Maxim machine guns.[32]

The Boers commenced their offensive with a invasion of the Cape Colony and Natal and took up positions, deep into British territory, in preparation for the British counter-attack. The tactics employed were those that had gained them success in the first Anglo-Boer war, and their expectation was for a second quick win, and to this extent they had little appetite in taking the strategic seaports, of Durban and Cape Town as planned. During this early phase the Boers had a numerical advantage, of 3 to 1, and could have been more decisive, with better leadership.

The war commenced on 11 October 1899 and by November the Boers had completed their initial action and occupied areas inside the Cape Colony and Natal, and had further laid siege to the towns of Ladysmith, Mafeking, and Kimberley. Key strategic points were also occupied, such as the towns of Colesberg, Arundel, Barkley East, the rail junction at Stormberg, and the Tugela River crossing at Colenso in Natal.

Although the British had been mindful of Boer posturing and re-arming over the previous three years, the British Army had done little to prepare for a full-scale conflict. The Army been scaled for small colonial campaigns and financially limited for preparing for anything greater. At the start of hostilities there were only 22,700 British soldiers in or on their way, by sea, to South Africa. Although acting offensively the Boers had yet again exercised caution and limited their gains and advances; had they been more ambitious they might have gained a strategic advantage, with the 3 to 1 ratio of force, stacked in their favour. The initial Boer gains were held and deep inside the British territories, but Boer looting and sacking allowed British troop to disembark and deploy reinforcements from Cape Town and Durban, and open a full-scale conventional conflict on two fronts; one through Natal and one through the Cape Colony.

By the end of November 1899 the British were confident in the knowledge that all that was required for the Natal Front was a single Army Corps of 46,000 men and materiel under the command of General Redvers Buller VC. His initial plan had been to concentrate all his forces just south of the Orange Free State border, cross the Orange River and push up the rail line, capturing Bloemfontein and Pretoria in one swift move. This plan was however thwarted by the sieges and he was forced then to divide his force into four to tackle the two towns of Ladysmith and Kimberley and also to have two columns head for Colesberg and Stormberg.

*A typical Boer commando: 'just farmers', in British eyes.
In battle they proved to be highly accurate riflemen, great horsemen,
adept in camouflage, fieldcraft and living off the land, who had
a determination to defend their homeland.
Courtesy of Ditsong National Museum of Military History Archive.*

This strategy however, was the undoing of the British offence – and of Buller, previously known for his outstanding battlefield results. In what became known as 'Black Week', three British defeats sent shock waves across the British Empire. In three key battles, Stormberg (10 December 1899), Magersfontein (11 December 1899) and Colenso (15 December 1899), Buller, who was in command, became known for shameful reverses and for committing the ultimate military sin, of losing the guns. Yet again the Boers failed to fully capitalise on their successes and instead of pressing the British south, allowed them to regroup and reinforce and for a counter-offensive.

After this shameful week the British government replaced General Buller by bringing Lord Roberts out of retirement to assume command in South Africa, with him Lord Kitchener as his chief-of-staff, the finest the British Empire could muster to resolve a deepening crisis. They arrived in theatre on 10 January 1900, and quickly set about planning a new more consolidated offensive, based on Buller's original plans, while he continued in vain at the task of liberating Ladysmith. This time Lord Roberts had arrived on South African soil and the war was not over, he was ready to defeat the Boers and avenge Majuba!

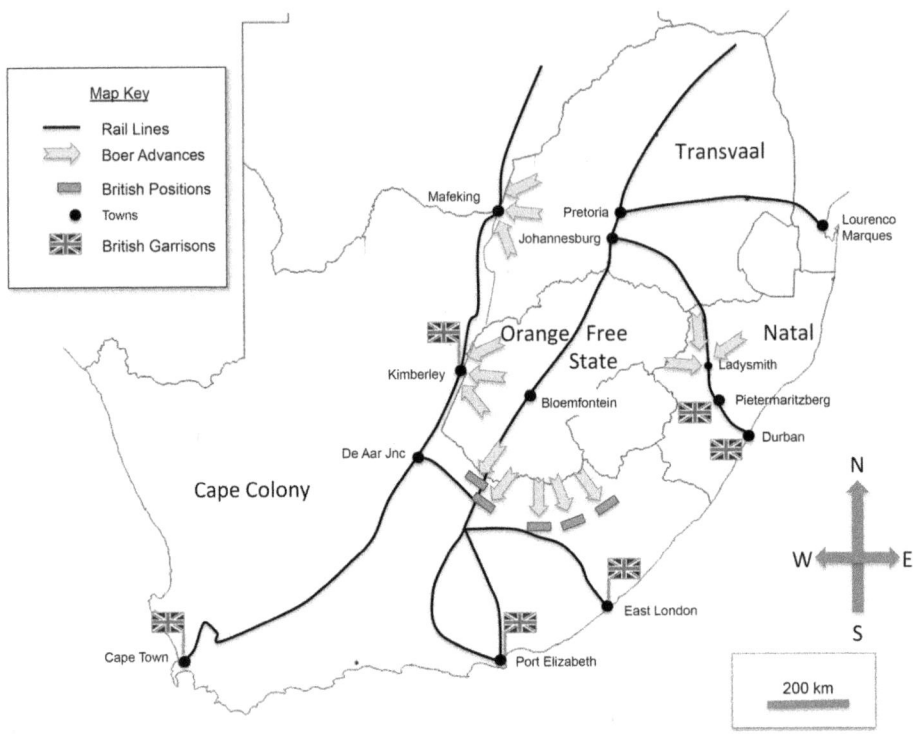

Initial Boer and British positions in 1899.[33]
Courtesy of André Wessels.

Phase 2: British Relief and Occupation

By 11 February Roberts was ready to launch his offensive and with 50,000 men and 110 artillery pieces managed to outflank the strong Boer position at Magersfontein, allowing General French's cavalry to sweep into Kimberley on 15 February 1900 to much acclaim, after 124 days of siege. This was quickly followed up with a significant success at the Battle of Paardeberg (17–27 February 1900), when General Piet Cronje was caught with his large commando and wagon train part way through a river crossing. The subsequent encirclement and bombardment inflicted heavy losses on the Boers, and over 4,000 were defeated and captured. The conventional fighting was beginning to turn in favour of the British.

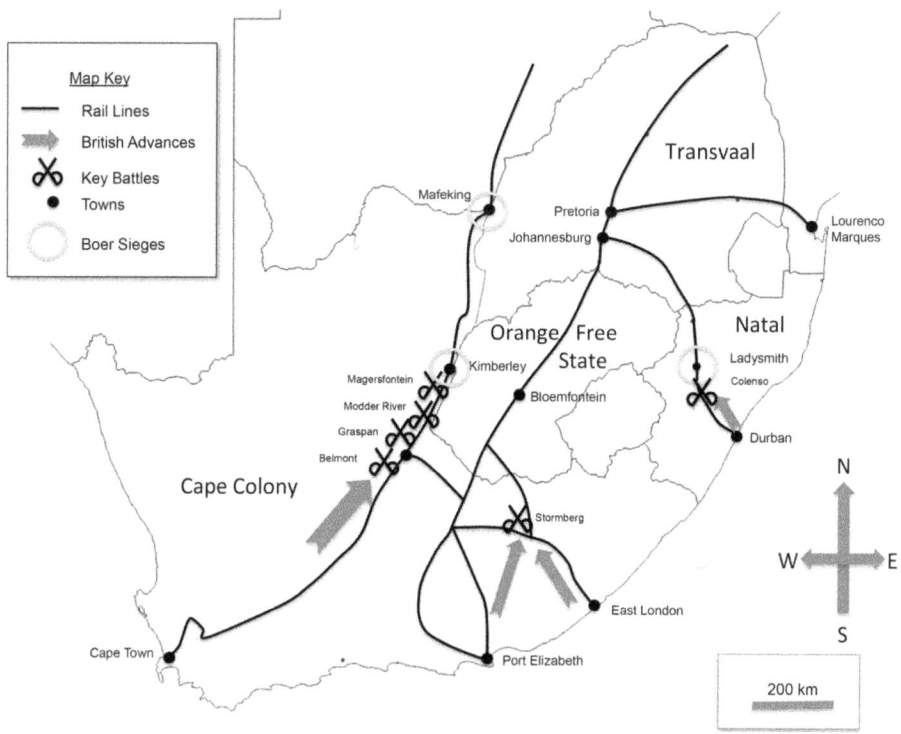

The British offensive to December 1899.[34]
Courtesy of André Wessels.

Boer defeats continued into March at the Battles of Poplar Grove (7 March 1900) and Driefontein (10 March 1900) before the Orange Free State capital of Bloemfontein was captured without resistance on 13 March 1900. This rapid advance had restored British pride and confidence, but it came at a price. The long and speedy marches had taken a heavy toll on the health and well being of his men and an epidemic of enteric or typhoid quickly broke out and spread throughout the force. Roberts was forced to pause in his drive northward to Pretoria for seven weeks, in order to regroup and recover. Meanwhile on the second front in Natal, Buller had taken heart from Roberts's success and Boer waning morale after the Battle of Paardeberg and by 27 February 1900 managed to relieve Ladysmith after a 118-day siege.

It was during this pause that the State Presidents of the two Republics met at Kroonstad for a council of war (*krijgsraad*) to decide their next move. Boer resolve stayed strong: they would continue to fight, but with new tactics. No longer could they fight the larger British Army units in defensive battles to stop their advance north. The decision was made to fight a less conventional war by fighting in smaller more mobile units focused on a harassing engagement based on guerrilla tactics. Although the terms had connotations of banditry the older Boer leaders did not appreciate, the mantle for this type of warfare was quickly and readily accepted by the younger leaders. Generals such as De Wet, Botha, Smuts and De la Rey were quick to act and adopt this more mobile approach and cast the die for nearly another two years of guerrilla conflict. By exploiting mobility and surprise they were able to exact maximum damage on the British for minimal losses, which is key when a force is numerically inferior. The failure to exploit this advantage in mobility and parity of numbers at the start of the war was identified by Deneys Reitz as follows: 'We did not know how our strength and enthusiasm were to be frittered away in meaningless sieges, and in the holding of useless positions, when our only salvation lay in rapid advance.'[35]

Quickly after the decision to adopt guerrilla tactics, De Wet in his own inimitable style decided to take the British garrison of 200 men at the Battle of Sanna's Post (31 March 1900) with a force of 1,500 burghers. On realising a much larger column was approaching from the west, the master tactician reorganised his force to ambush the larger column of 1,800 to achieve a classic surprise attack. It was a crushing defeat for the British with over 140 casualties and 426 prisoners of war taken with minimal Boer losses. In addition seven artillery pieces and over 100 wagons were taken for much needed resupply, and the template set for the future of many Boer attacks. The legend of De Wet had been created.

By now Roberts was ready to resume his advance towards Pretoria and he set out with 26,000 men on 3 May 1900, quickly advancing to reach the Vaal River on 28 May. Now that the Orange Free State was annexed and renamed the Orange River Colony he completed the short distance to Pretoria in another eight days, marching in unopposed on 5 June 1900. As both Republican capitals, the city of Johannesburg and the goldfields were under British control, it was Roberts's beliefs that the Boer will to continue would wane and victory would be won soon. The siege of Mafeking did not end until 17 May 1900, after 217 days of siege for which Robert Baden-Powell – later founder of the Scout movement – became a national hero.

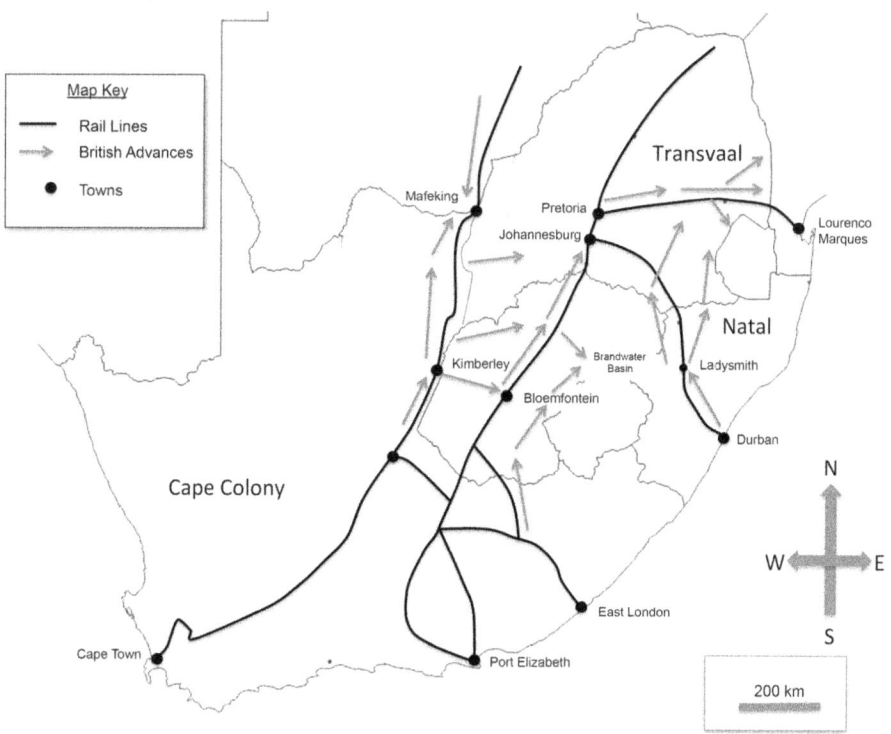

The British offensive from February to September 1900, including relief of the three siege towns.[36] *Courtesy of André Wessels.*

With De Wet very much still on the loose and active, Kitchener was dispatched back to the Orange River Colony to oversee repairs to the rail lines and capture the elusive Boer commander. Kitchener failed to capture De Wet and came very close to being caught himself one night in his pyjamas, having to flee to a nearby unit for safety. De Wet was still free to conduct operations, but nearly became trapped in the Brandwater Basin, where a large force of Boers was captured at Surrender Hill (28–30 July 1900). The Boers under Commandant Prinsloo surrendered without a fight, losing over 4,300 men, three guns, 2 million rounds of ammunition and over 6,000 horses.[37] This marked the end of the Orange Free State as a sovereign entity and any conventional resistance to British rule.

Phase 3: The Guerrilla War

The guerrilla must move amongst the people,
as a fish swims in the sea.
MAO ZEDONG

Once Pretoria had been secured Roberts moved eastwards with 20,000 men towards Komatipoort in order to secure the railway line to Lourenço Marques (now Maputo), defeating the Boers at Bergendal (20–27 August 1900). On the Natal front, Buller coordinated with Roberts's advance, aiming to catch the Boers in a pincer movement near Machadodorp. Buller attacked in the centre of a long defensive line, catching the Boer General Louis Botha by surprise. This weakened the Boers' position, making it untenable, and the British were able to gain ground while the Boers retreated. The ZAR's seat of government, vacated from Pretoria, was now on a train at Machadodorp, and President Paul Kruger now had to retreat still further east to Nelspruit. This was the last time the Boers were to attempt to fight a defensive pitched battle; the guerrilla phase was now fully in progress.

Roberts now claimed the former ZAR, renaming it the Transvaal Colony. He considered his task of defeating the Boers had been achieved by seizing their capitals; mopping-up and police work were all that remained to finish off the war. He departed for England in late November 1900, handing over the supreme command to Kitchener. Roberts had certainly turned the tide in favour of the British, but he had not decisively defeated the Boer forces; large numbers of men and materiel were still intact and free to conduct operations. He had started limited Boer farm burnings as reprisal for rail sabotage by the commandos, and laid the foundations for a deep rooted and determined guerrilla war. The British were only in control of key towns and supply lines; small towns and vast areas of veld were uncontrolled as yet.

This 'final phase' was to last nearly two years. Train attacks on the lines in the former Republics were now being prevented with patrolling and the advent of the blockhouse. Kruger had departed to Europe, leaving the younger Boer leaders to continue the fight, opting for 'hit-and-run' guerrilla tactics. One of the Boers' key aims by protracting the war was to force the British to negotiate a peace and gain a more favourable compromise. The war was beginning to drag; war weariness and the prospect of a new British government with the Liberals in control might set the conditions for the Boers to gain this more peaceful outcome.[38] In addition, foreign powers were also there to exert pressure on the British and provide support to the Boers where possible. With the conservative and cautious Kruger out of the country, the more dynamic and bold young leaders in charge of the commandos set out to show the British that they were not beaten. The guerrilla phase was keenly followed by Queen Victoria, who described it as a 'very difficult, anxious and troublesome guerrilla war'.[39]

Roving commandos now turned their attention to the Cape Colony, deciding to take the fight back to the British in their own territories and with the hope of drawing more Cape Afrikaners into the war. About 11,000 had initially joined the Boers, but subsequently only another 5,000 took up arms and a full-scale rebellion never materialised.[40] Boer commanders such as De Wet, Scheepers, Fouche, Malan and Maritz conducted wide-scale operations against the British in the Cape Colony, but significant 'blockhousing' of the Colony's railway infrastructure prevented any major acts of sabotage.

At the start of the guerrilla phase Kitchener had over 200,000 men in theatre, which at least on paper gave him overwhelming superiority over what was estimated to be a 10,000 strong Boer force. These figures are however somewhat skewed at face value: Kitchener had half his force tied up in lines of communication and garrison guard duty and the Boer forces in the field, as estimated by historians, were closer to 30,000 or 35,000.[41] A ratio of only 3:1 against a highly mobile, agile unit with keen local and terrain knowledge set against an Army lacking mobility and intelligence, was not a good start point with which to sweep the Boers off the veld.

The Boers had tremendous advantages over the British and were very successful in evading their large roving columns and static blockhouse lines. They capitalised on their local knowledge of the terrain, were provided with good intelligence, kept mobile as a key form of defence and were provided with inspirational leadership based on a heartfelt desire to retain their land. The Victorian Era small wars and colonial campaigns had not prepared the British politicians, Army Command or its field generals for this type of determined and sustained resistance. New ways to fight would have to be found.

British colonial experience, had taught them that the civilian population was a liability for an enemy fighting on its own soil against an outside force. In the Second Anglo-Afghan War (1879–1880) Roberts[42] had ordered villages to be burned to the ground, and he had already started punitive reprisals on Boer farms for the commandos' acts of train sabotage. The farms were a known source of supplies for the commandos and many Boer men were allowed to return during the early stages to resupply allowing them to prolong their time 'on commando'; now the farms were to become 'legitimate' military targets. Removal of this vital supporting infrastructure, started by Roberts, was now developed into a full scorched earth policy by Lord Kitchener: farms were burned, stock removed, and inhabitants cast onto the veld or 'interned' in what would become known as the first 'Concentration Camps'

The British and Imperial soldier,[43] many confined to blockhouses, the remainder on horseback in the guerrilla phase hunting down the Boers. They were highly regimented, in a strict chain-of-command and adept only at fighting in fixed formations commanded by officers.

The land was to be stripped of all its support, laying waste to large areas of the Orange River Colony and Transvaal, destroying up to a maximim of 30,000 Boer farms and homesteads, and very many more other buildings – including all those owned by black farm workers. White and black families then became the burden of the ill-prepared military and civilian administration: over 67 white and black camps[44] were hastily erected near railways and towns across the country. Although denying vital support to the commandos as it was intended to, it was also counter-productive in other areas. There was a huge negative publicity backlash in Britain, which questioned the war effort and damaged British reputation. It further garnered support for the Boer war effort from sympathetic nations and considerably hardened the Boer resolve in the field to continue to fight. 'Go and fight, I can get another husband but not another Free State,'[45] as one wife was reported to remark. Militarily, however, the combination of Scorched Earth and forced removals did denude the veld of infrastructure and weaken the commandos' combat capability.

The next step was to destroy the fighting capacity still loose on the veld. There could be no victory without a defeated or captured enemy; the Boers had not yet been driven to the point of surrender. The war was now fought on the open vastness of the veld on which the commandos roamed. Regiments and battalions were now formed into manoeuvre units called columns. They consisted of between 200 and 1,500 mounted troops and their immediate support, such as pom-poms and huge supporting ox-wagon trains. These were used both for the destruction of farms under Scorched Earth and in the more agile requirement to chase and catch commandos. By the end of the war over 90 columns were employed on these duties in the 'parcelled-up' areas of the veld. The development of these key tactics employed during this phase of the war is fully explored in the next chapter.

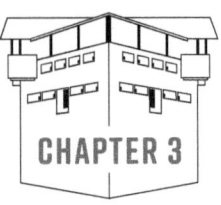

CHAPTER 3

BRITISH STRATEGY AND TACTICS

Strategy without tactics is the slowest route to victory.
Tactics without strategy is the noise before defeat.

SUN TZU, *THE ART OF WAR*

Two radically different fighting traditions clashed in the Anglo-Boer War. The British Army fought from its historical roots as a parade ground army led by socially superior generals, while the Boers fought as a citizen army, and latterly as mobile commandos using ground-breaking guerrilla tactics. It was hardly surprising therefore that the strategy and tactics used against the Boers changed drastically during the course of the war. The British were at the foot of a steep learning curve, unusually for the days of the Empire, fighting a non-native force, which used modern weapons and wide-ranging tactics. The war took place at the cusp of a new century and the beginning of a new era of warfare.

In this campaign the war was initially fought from south to north: from the British Cape Colony due south and Natal in the southeast, and into the two Boer Republics. Once the British counter-attack and invasion of the Republics had been achieved, the railway systems of both the Orange Free State and the Transvaal Republic became important as key supply routes or lines of communication. Thus railways from the coastal ports of Cape Town and Durban became the skeleton for the vital supply lines of communication and later the defensive backbone from which the blockhouse system grew until the end of the war.

The railway lines were already protected by barbed wire,[46] but only to prevent animals crossing in front of trains. For military protection wire alone was useless; the universal military mix of obstacles and covering fire would be needed for initial protection of these vital supply routes.

BRITISH STRATEGY AND TACTICS

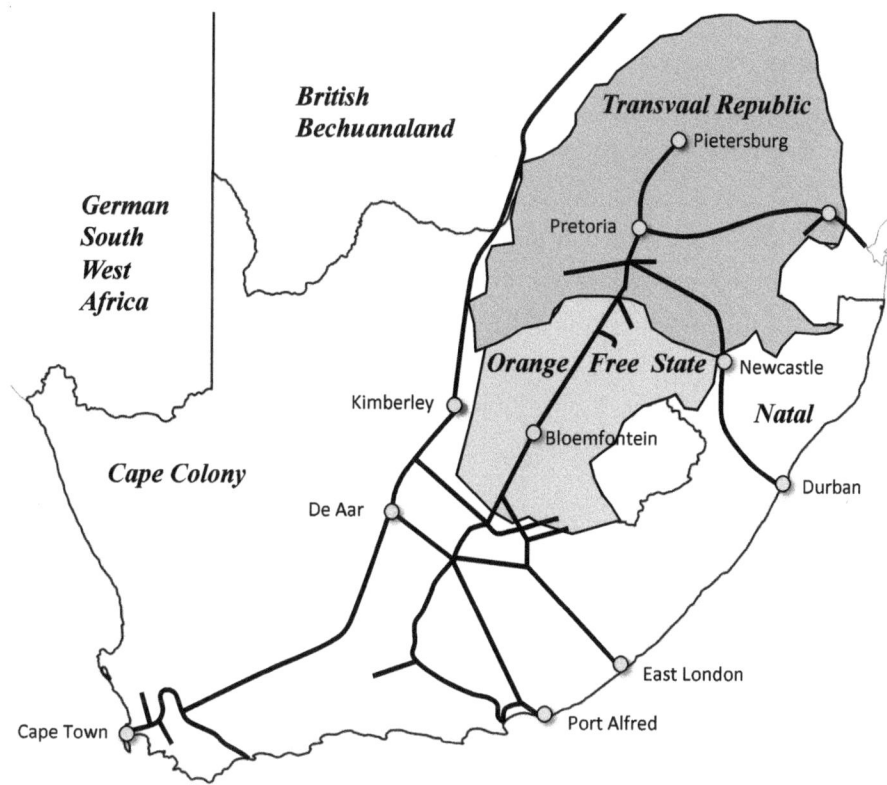

Railway lines in 1899 prior to the start of war.[47]

After the initial Boer attacks into the Cape Colony and natal; the dreadful losses of Black Week; and the lengthy three sieges on the towns of Ladysmith, Kimberley and Mafeking, the British had to organise and settle into a more measured campaign to defeat the Boer Republics. This was to be no 'quick win' campaign as had been imagined by the London War Office. The relief of the siege towns allowed the British, under Lord Roberts, a platform from which to make advances towards the cities of Bloemfontein, Johannesburg and Pretoria in the heart of both Boer Republics. It was assumed that when these towns fell to the British the war would be over.

COUNTERING THE TRAIN ATTACKS

Once Field Marshal Lord Roberts had captured Pretoria, in June 1900, he quickly sought to secure both the southern long lines of communication back to the Cape Colony and Durban but also eastwards to the Delagoa Bay seaport. British forces moved out of Pretoria in July to repair and secure this important strategic rail link for much shorter access to a seaport. By August, work was completed as far as Komatipoort, the easternmost station in the Transvaal and on the border with Portuguese East Africa.[48] In addition to the repairs there were also troop movement facilities installed to fit the line for military use according to the geography and nature of the ground, such as additional stations, water supplies and electric lights. The line was finally ready for military use in late September 1900.[49]

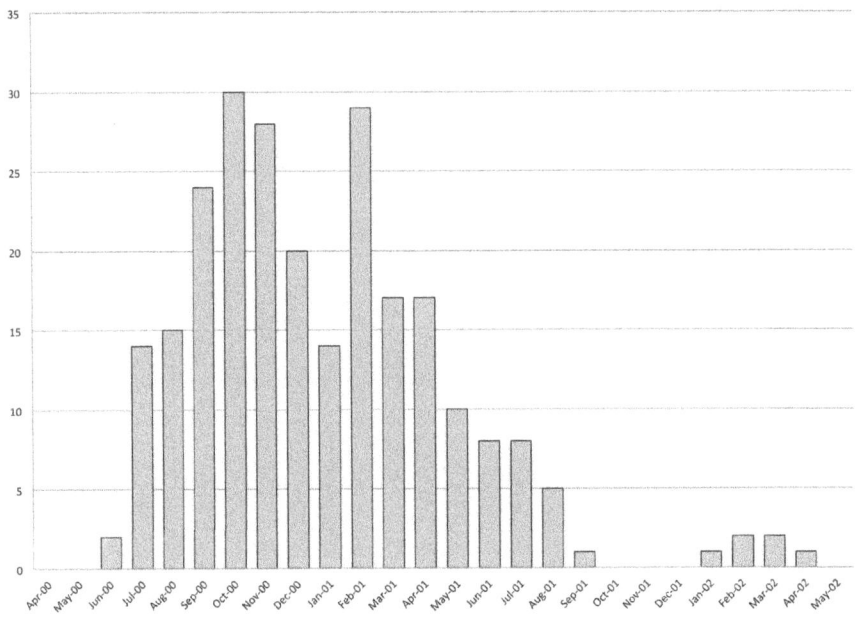

Frequency of train attacks.[50]

Up to this point the Boers had fought a conventional type of war, but by avoiding being drawn onto a large fixed battlefield. The British had expected them to surrender once the key towns were captured; however Boer resolve and determination had been underestimated. Boer commanders elected to move into the field and in their commando structures conducted mobile operations that caught the British completely off guard. Up to this point the Boer commandos were still largely intact and operational in the area of the Transvaal, and were prevalent in attacking the troop trains now returning the construction force back to Pretoria. During September to October 1900, 102 troop trains conducted this work under Boer fire by day and night.[51] Train attacks, which peaked at 30 per month during October 1900 and February 1901,[52] cost the British heavily in terms of men and materiel as well as significantly delaying the plan to dominate the two Republics. The resolve and determination of the Boers to defend their territory by whatever means available to them was to be a thorn in the British side up to the end of the war.

Jack Hindon[53]

Jack Hindon.[54]

The most famous of the train wreckers was Jack Hindon, who is buried in Middelburg Cemetery. Known as 'Dynamite Jack', Captain Oliver Jack Hindon was Scottish by birth and arrived in Natal after joining the British Army as a boy. He deserted, claiming that he had been physically assaulted by members of his unit, and settled in the Transvaal Republic, becoming a stonemason and later a police officer. After fighting against the British during the infamous Jameson Raid, for his loyalty he was awarded citizenship of the Boer republic, an honour not usually given to an *uitlander*. At the outbreak of war Hindon deployed with the Middelburg Commando, which fought with distinction, most notably at Spion Kop.

As the Boers took up guerrilla warfare, Hindon formed a unit known as the Hindon Scouts. The unit was best known for its train-wrecking operations along the Pretoria-Delagoa Bay Railway Line, operating under General Ben Viljoen. Hindon was relentless in the early phase in attacking the railways, but his activities were eventually curtailed by the deployment of blockhouses and an overwhelming superiority of troop numbers protecting the line.

Hindon's sabotage activities were well ahead of their time, being some of the first to use improvised explosive devices on the battlefield. Using the firing mechanism of a British Martini-Henry rifle and basically a bucket full of dynamite, the charge would be initiated by a passing train, to devastating effect. The derailed train would then be looted for supplies. Kitchener remarked that Hindon had caused 'more difficulties for the British and Colonial forces than any other Boer Commando'.[55]

From 1970 to 1975 the Republic of South Africa instituted a medal called the Jack Hindon Medal (JHM), which was awarded to other ranks for exceptionally diligent service in the Commandos, which had by this time morphed into a rural defence component of the South African Defence Force.

SECURING THE KEY POINTS

Roberts's initial defences along these important lines of communication during the latter half of 1900 were rudimentary in nature, consisting of trenches, sangars and sandbagged positions. Key points were identified as those most vulnerable to Boer attack, such as bridges and drainage culverts (where charges could be placed to destroy or significantly damage single-span or multiple-span crossings of rivers) and strategic points such as rail-heads, staging areas and military and civil stations. The deployment was largely ineffective, requiring a large number of troops to both occupy the trenches and patrol the rail tracks. The rail losses were becoming costly in terms of materiel and men, and it was clear a much more effective solution had to be found.

Early fortified house with trench and wire protection at Krugersdorp Station.[56]
Courtesy SANDF Photo Archives (Pretoria).

Roberts was now acutely aware of the need for more robust defences and it was in September 1900, on his return from a trip to Komatipoort, that the first use of the term 'blockhouse' was recorded. The forces of the period were quite well acquainted with the use and deployment of blockhouses, which had been widely used in several theatres prior to the South Africa operations. It was Kitchener (also an Royal Engineer) who on assumption of the South African command in late 1900, sought to upgrade the defences in order to more effectively protect the long lines of communication. The work of the

Engineer-in-Chief, Major-General Elliot Wood, in designing the Elliot Wood Pattern stone blockhouse was key to this early deployment and successful use of blockhouses. They were however prohibitively expensive, took months to construct and although fit for purpose for key point defence, only very few were initially built during early 1901 to protect the Cape Colony railway network.

ESTABLISHING A CONSTABULARY FORCE

In mid-1901, in an attempt to increase the force levels on the ground, a new unit was fielded, and called the South African Constabulary (SAC). Under General Baden-Powell approximately 10,000 troops were raised, in the main part from Great Britain, with some also recruited from the colonies. He selected men carefully and set about training them for two months with the primary aim of policing the two former Boer republics. Initially however they were to be used in a more aggressive role and were supplementary to the blockhouse lines. Although well recruited, their effectiveness in theatre was questioned by Ian Hamilton,[57] who wrote in late 1901: 'The SAC are thought to be the finest body of high class soldier England has ever sent out, but sadly miserably handicapped by lack of administration capability shown by BP (Baden-Powell) and Ridley.'

A number of SAC posts and lines were built in the Transvaal and Orange Free State. The first chain of posts was established in the Free State from Bloemfontein to Petrusburg. Although termed 'posts' they were more fort-like in terms of size, the strength of force they could house and the duties they were intended to fulfil. Duties were torn between civil support in a policing role to suppress civil unrest in the British-controlled areas supported by Milner and Baden-Powell, and Kitchener's original concept of discharging military duties. As they were set quite far apart, there were instances when intermediate blockhouses were built in between the SAC posts to assist in providing the linear coverage required.

The deployment of the blockhouse lines at this time was still in the realm of static defence for the rail lines and not for any other use. Also around this time, General Bruce Hamilton was sweeping a section of countryside with seven small columns in an attempt to catch Boers. The tactics of these early drives relied on pace to try and overtake the Boers; this time, however, something new was tried. In his plan the line of SAC posts between Bloemfontein and

Petrusburg was to be used as a wall against which the Boers were to be driven. Between 5 and 8 June 1901 his sweep of the veld onto the Constabulary line captured 268 enemy prisoners, and was extremely successful compared to previous drives.

It seems this may have been the key to commencing the cross-country blockhouse lines. Kitchener perceived that if these relatively loose chains of police posts had rather been blockhouse lines with fortifications set more closely together, then success would have been much higher. In concentrating the blockhouses at distances much less than the 5 miles (8km) for the SAC posts, the net would be so much tighter and harder to breach. The first of these lines were deployed east from Bloemfontein to Ladybrand, and then west to Jacobsdal, and were completed by December 1901. In the Transvaal, lines around Pretoria and Johannesburg followed quickly after, and the final carving up of the veld with dedicated cross-country lines commenced in early 1902 to the end of the war in May.

PROTECTING THE KEY TOWNS

After the British had taken the key centres of population it was important to ensure that they were secure and well defended from counter-attack. Once taken, Roberts believed this would be the strategic turning point that would force the Boers into surrender, therefore their continued occupation was considered vital. Both the former capitals – Bloemfontein for the Orannge Free State, and Pretoria for the Transvaal – had to be made into protected areas with inner and outer lines of defence. The area surrounding Johannesburg was also vital, as the gold-producing mines were located on the Witwatersrand ridge running through the city from east to west.

Roberts captured Bloemfontein on 13 March 1900, but despite this there were engagements around the town at Karee Station (29 March 1900) to the north and at Leeuwkop (22 April 1900) to the southeast, during the period Robert's troops were resting in the town. By early June Johannesburg and Pretoria were captured and in July Buller's Natal advance joined up with Roberts. The capture of the key capitals was completed. Bloemfontein's defences however in terms of blockhouses and SAC posts surrounding the town were not formally completed until June the following year. In October 1900 Major ED Haggitt took over command of 7th (Field) Company, which was responsible for these defences. They spent the next few months remodelling the town's defences to suit the reduced garrison and the absence of enemy artillery. By April 1901 the original entrenchments had been replaced by 30 closed and self-contained works of various types, linked by wire entanglements.[58] Once finished the town was protected by an ever-deepening layer of defences, with the town centre fortified in terms of all the entry and exit points. The Bloemfontein War Museum, Rice Pattern blockhouse would have been a part of this defence before its relocation to the museum grounds.

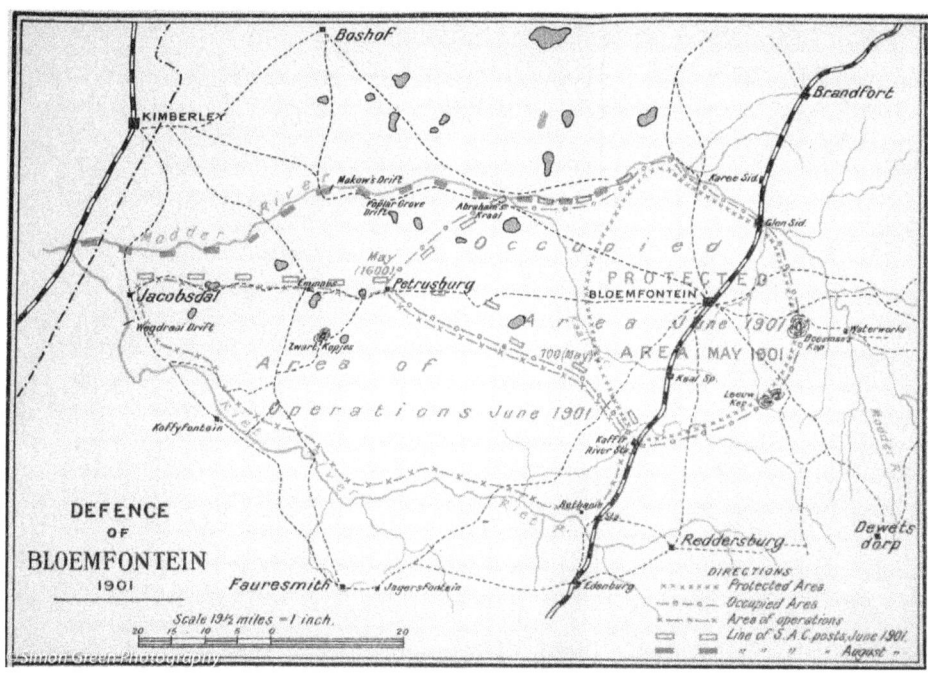

The defence of Bloemfontein by August 1901.[59]

The town was further cocooned in an outer protected area, a 20 mile (32km) radius in all directions. To the west, as shown in the map, there was a secondary Occupied Area, bounded in the north by the Modder River; from Abraham's Kraal, through Petrusburg and then south-east to Kaffir [sic] Station (Riversford). Even further westwards there was a larger area of operations, reaching in an arc as far as Jacobsdal and bounded by the Modder River's SAC posts to the north and the Riet River to the south.

Work on Pretoria's defences commenced immediately after its occupation in June 1990, initially on a temporary basis with ditches and sandbagged positions. The Commander Royal Engineers Pretoria District records that

Johnston's Redoubt, East Fort, and Howitzer Redoubt were 'well advanced' by 1 December 1900.[60] This type of defence for the capital had started well before the Elliot Wood and Rice Pattern blockhouses had been developed, but marked the beginning of the fortification of the three key towns. As these developed Pretoria was surrounded by a ring of blockhouses, forts and redoubts, in order to ensure its continued security. The former capital was also the hub of the rail lines, north to Pietersburg, east to Komatipoort, west along the ridge into the Magaliesberg and south to Johannesburg and the seaports.

Johannesburg's defences did include blockhouses and there are references to blockhouses mainly on the key roads in and out of the city and on the prominent ridges, but sadly none have survived 20th century urbanisation. This city's key assets however were the gold mines, on the East Rand towards Springs and on the West Rand out towards Klerksdorp. The defences consisted of single point protection for some of the key engineering assets such as the headgears, localised patrolling by units such as the Rand Rifles, and lines of blockhouses into the western Transvaal securing the whole mining district.

*Blockhouse on overhead staging at a gold mine near Johannesburg.[61]
Reproduced by permission of Durham University Library and Collections.*

SECURING THE LINES OF COMMUNICATION

The Royal Engineers were also giving the issue of train attacks some thought and Major Spring Rice RE, Officer Commanding the 23rd Engineer Company based in Middelberg, had been experimenting with cavity-walled structures. The first octagonal version blockhouse was large and cumbersome to erect but provided good all-round defence from the eight-sided loopholed walls. Early versions built at crossings over the Klein Olifants River, close to Middelberg, resemble the ones still standing in Fouriesberg and Pretoria. An octagonal second version quickly followed which was lighter and quicker to assemble, but having to bolt eight sides together made it an ineffective design.

By December 1900, the British were occupying the Transvaal and Free State, and Lord Roberts was on the boat home to England, having handed over to his recently promoted deputy Lord Kitchener. The war was considered to be over; Kitchener only anticipated adminstrative duties to be completed and consolidated. The Boer commandos however had other ideas, and merely holding and controlling the cities would not be enough to defeat these men of the open veld. The Boers commanded all the skills to conduct a guerrilla war and would remain an effective fighting force against the British for another 18 months. The train attacks continued throughout the area and the threat of counter-attack into the Cape Colony and influence over the Cape Dutch community to come over to the Boer Republic side remained a constant concern. Kitchener now had an all-out guerrilla war to fight across a vast expanse of open veld scattered with Boer family farms and small sympathetic towns.

Elliot Wood had convinced Kitchener to commence work on truly fortifying the rail lines with masonry blockhouses, and had designed a two-storey model. This is accepted as the Elliot Wood Pattern or Standard Pattern design blockhouse and remains the most well-preserved type of its kind across South Africa.

Elliot Wood blockhouse, Merriman, Karoo.

Twenty-five examples – deployed predominantly along the rail line from Cape Town to secure the main bridges spanning significant rivers – exist in various states of repair, from the pristine overnight stay blockhouse at Stormberg Junction to the ruined Leeu Gamka, with the most northerly remaining examples of this type at Modder River and Harrismith. Many were deployed in pairs, such as those remaining at the remote Dwyka River, Beaufort West, Laingsburg and Wellington. The main build period for this phase of blockhouse construction was between January and July 1901, but most likely continued after that period with continuous enhancements.

The Elliot Wood Pattern blockhouses stand over 120 years later as testament to a good design and build of exceptional quality for a field defence. At the time, however, this came at a price; they took about 30 artisans three months to complete, requiring locally quarried and cut stonework, and cost approximately £800 in material alone. For the purpose of securing strategic bridges they undoubtedly served their purpose, but to secure thousands of miles of rail lines, cuttings and culverts a quicker and more cost-effective solution was required.

Rice Pattern replica blockhouse, Carnarvon.

By using the pre-fabricated corrugated iron blockhouse developed in January to March 1901 by Major Spring Rice to fill in the lengthy gaps between the key bridges where the Elliot Wood Patterns were situated, the whole rail line of communication was fully secured by the last quarter of 1901. These Rice Pattern blockhouses were mass-produced in their thousands as small temporary strongpoints, few of which remain intact today.

Once the area and railway lines were secured there were 20 armoured trains[62] in operation to supplement the blockhouse lines and provide rapid reaction reinforcement should Boer attacks or crossings occur. Operating from sidings, they patrolled the rail lines and acted as additional support once the drives to round up the Boer commandos commenced. The Boer train attacks had steadily dwindled to only a single incident per month by September 1901.[63]

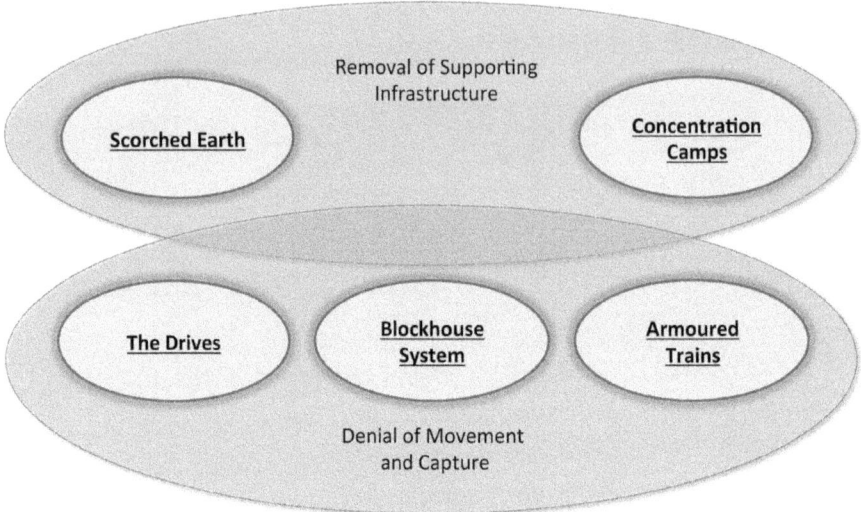

The components of Kitchener's strategy.

SCORCHED EARTH POLICY: REMOVING THE COMMANDO SUPPORT STRUCTURE

After the capture of Pretoria in mid-1900 Roberts started to implement a number of tactics to try to avert the train attacks. One of the first was to place notable Boer civilians on the footplates of the trains, as hostages to provide surety to the train and in order to protect the driver from sniping and the train being derailed.[64] This was not an extraordinary practice: it had already been used in the Franco-Prussian War (1870–1871) and in the American Civil War,[65] and in a later case local Ministers were also used. Today they would be viewed as 'human shields' such as used by Saddam Hussein in the Iraq war of 2003 and their use highly controversial. For the British it proved to be largely ineffective, and eventually the civilian authorities did wake up to the fact that this was neither ethical nor effective. The next step was to raze to the ground any Boer farm that was in close proximity to the train attack, the aim being to deter future attacks, which was again ineffective as a deterrence.

The policy of scorched earth was thus born in late 1900, and Boer farms were often looted, burnt and then dynamited, cattle taken or more often killed on the veld, and families (mainly elderly men, women and children) relocated in refugee camps. These camps eventually became known as 'concentration camps' and the term was coined in the popular press and military history for the first time in English, although it had been previously used in Cuba in 1895. In addition all crops were destroyed, so that the Transvaal and the Orange Free State effectively became barren wastelands. Black workers were also similarly displaced but history has been slow both to acknowledge their plight and to monumentalise their passing. Only since 1994 did the black camps and deaths start to be memorialised. It was a distasteful concept, objected to by Milner, and loathed by generals and soldiers alike who had to execute the policy dynamiting farms and turning women and children onto the veld, often with only a few minutes' notice.

Initial reprisal farm burning was agreed on by Roberts and Kitchener on 27 August 1900 after the Battle of Bergendal, and related not only to farm attacks but also to telegraph line cutting. Ian Hamilton recalled the decision and considered it ill-advised, and that the tactics previously adopted in Kandahar by Roberts, and in Khartoum by Kitchener, were not suitable against white men who were more attached to their homesteads. By the end of 1900 Roberts had left Kitchener to 'mop-up', but events were not going to plan. Kitchener took the reprisal burning policy started by Roberts into a systematic destruction of the Boer farms across both Republics, in order to deny the commandos a sympathetic and sustaining infrastructure as the Boer commanders opted to fight a guerrilla war.

One of the first farms to be destroyed was that of Christiaan de Wet, the high-profile Boer general, but it was not just individual farms that were burnt. Often the farms and countryside within 10 miles (16 kilometres) were also cleared of farmers, stock and harvest. It was not only as a result of specific attacks on trains that this was implemented, and commanders on the ground had significant latitude to exercise discretion, both for those loyal to the Crown and those merely suspected of collusion. In many instances 'suspicion' related to harbouring 'spies' or scouts, withholding information, housing or providing supplies for commandos.[66] In some instances, although specifically banned, families or relatives of men on commando also had their farms burnt.[67] Additionally there was also a system of fines and levies put in place to further punish farm owners, but this does not seem to have been implemented, although De Wet claimed he was fined. The Roberts phase had destroyed as many as 630 farms from June 1900 to December 1900 as punitive reprisals, whereas Kitchener was to destroy 30,000 farms[68] in a systematic plan of destruction.

This attempt to force the Boers into surrender had the reverse effect, making the mobile and agile commandos more determined and resolute to fight on to the bitter end. Those diehards who continued to fight actually became known as the 'bitterenders' *(bittereinders)*. The war for Kitchener however was dragging and the Boer commandos were still an effective fighting force, which had to be crushed in order to end the war, so he continued what Roberts had started.

During this policy over 100,000 people lost their homes and were turned onto the open veld. The concentration camps, started as legitimate refugee camps, were now becoming fuller every month. Kitchener was determined to finish the war and to take any means he could to achieve this, desperate as he was to get to India for his next assignment. His resolution can be noted from Army Circular No. 29 which spelled out the actions in no uncertain terms:

Of the various methods suggested for the accomplishment of this object, one that has been strongly recommended, and has lately been successfully tried on a small scale, is the removal of all men, women and children and natives from the districts which the enemy persistently occupy. This course has been pointed out ... as the most effective method of limiting the endurance of the Guerrillas... [69]

As the destruction continued, by winter 1901 there was barely a farm standing north of the Orange River, and all the families were either in the concentration camps or 'on commando' with their menfolk. The commandos now had to be self-sufficient; no more returning home on leave to get provisions; this support had all been removed by Kitchener. The commandos now started to live off the land and the larger and more readily provisioned British units. They used a couple of tactics; one was to follow a large column and scavenge what it dropped; the second was to attack columns solely for the purpose of gaining provisions and ammunition. Rarely during these operations or after a successful battle did the Boers take prisoners. Often they would relieve them of all their clothes, arms, ammunition, horses and supplies and cast them naked onto the veld. The Boers called this *uitskud* or literally 'shaking out'. The scorched earth policy was certainly beginning to take effect, wearing down the commandos' ability to fight and denting morale. This was another part of the developing strategy that hampered the Boer operations and over time slowly demoralised the men fighting a war, knowing their families were suffering, their farms gone and with no tangible end in sight, fighting an Empire determined to win at all costs.

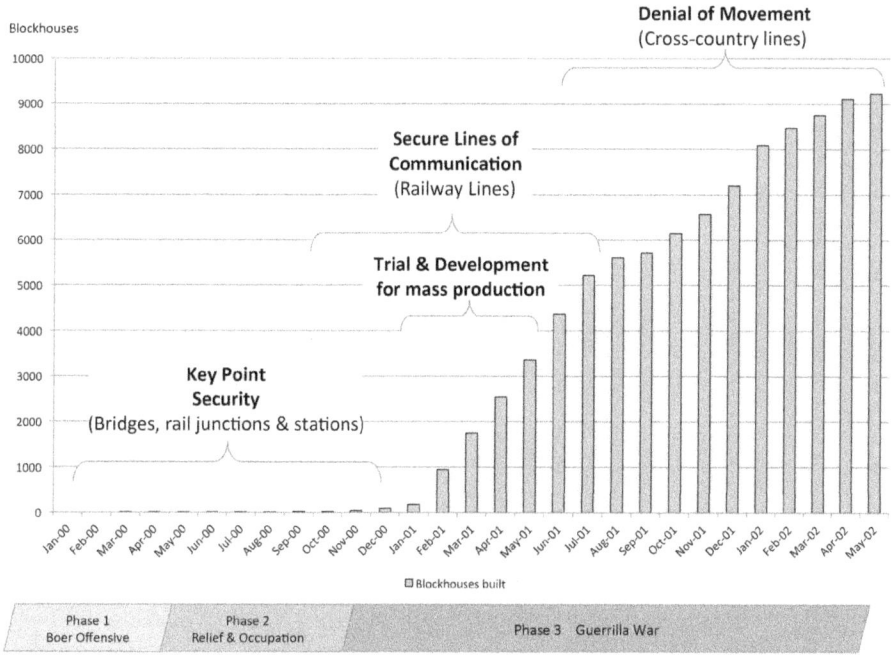

Estimated rate of blockhouse deployment.[70]

CONCENTRATION CAMPS

The first formal concentration camps were erected early in 1901. Thousands of men, women and children were removed from the farms and towns where they lived and placed in these camps. The concept of removing the supporting civilian infrastructure was not a new or particularly revolutionary one to the South African campaign. The policy had been used in a very similar manner in Cuba, along with blockhouses, as recently as 1896. The Captain General of Cuba had forcibly removed the inhabitants of the country's four western provinces into fortified areas – known as the re-concentration system or *reconcentrado* in Spanish –where they had died in many thousands. It is not known if Kitchener was aware of this; however, he was to become infamously associated with these internment camps as the so-called first 'concentration camps' in history. At the outset the British military authority was unprepared to accommodate thousands of people, and civilian authorities had made no strategic provision for food and medical supplies to the Boer men, women and children, let alone for the black camps. This resulted in malnutrition and disease, and the subsequent death of thousands of people in the camps.

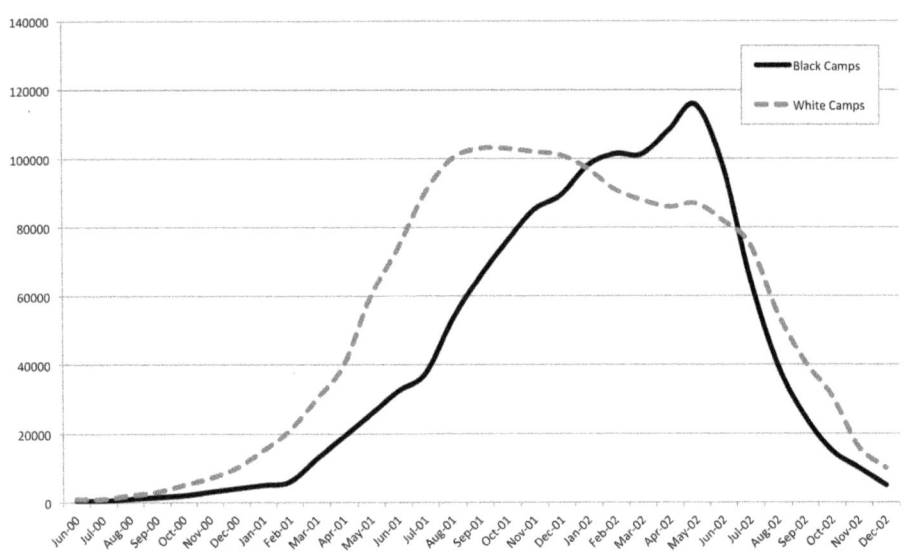

Concentration camp population in the Transvaal and Orange River Colony 1900–1902.[71]

That human suffering was an unfortunate necessity of war was not an uncommon concept for the British Army of the period, having been used to good effect in other small wars. Previously Lord Chelmsford's attempts to invade Zululand in 1878 had recognised the need to systematically attack the Zulu supply depots. This was in order to reduce the combat effectiveness of the Zulu regiments, weaken their ability to resist long-term conflict and reduce the King's authority. The latter two motivations attempted to undermine the will to continue by causing internal friction between the civil population and the military in the field, not dissimilar to that sought with the Boer fighters and their families in Kitchener's scorched earth policy.

The camps started under Roberts were temporary in nature and not designed for the widespread and lengthy internment of thousands of families. As the problem grew and the camps became both a logistical and political thorn in the side of the military, Kitchener off-loaded the responsibility to the civilian administration, which was equally ill-equipped in terms of funding or resources to manage the 42 white concentration camps[72] and between 68 and 89[73] black concentration camps. A number of the these camps were outside the Boer republics and others were temporary or transit camps, explaining why there were significantly more black than white camps, although the number of those interned were comparable, each about 110,000. Thousands of men, women and children died and this left a significant and indelible stain on Anglo-Afrikaans and Anglo-African relationships to this day.

The backlash for Kitchener from this policy was twofold. Firstly it hardened the Boers in the field to continue fighting to the bitter end, urged on by the women suffering in the camps, who often berated their husbands if they considered giving in. Secondly there was a huge outcry in Britain and Europe, which garnered even more resolute support for the Boer cause from Britain's rivals. The concentration camp conditions were getting ever more perilous, especially during the winter months, which exacerbated the death rates considerably.

It was not only the Boer families that were displaced, but also their *bywoner* or tenant farmers and black workers who also had to be sheltered and fed in camps. This required both white and black camps to be established, as there was an understanding that the two should not mix, although some Boer families took their black servants to their camps.

During 1900 very many more Africans were removed from the land than white families, though most probably never made it into a formal camp. To the British they were a cheap source of labour, to be offered a life in camps or work in the newly reopened gold mines. Simultaneous to the resumption of this economic activity was the establishment of the Department of Native Refugees in June 1901 under direct British military command. In addition, during the early phase, many local burghers fleeing the war came as people loyal to the British requiring to be housed and fed. They were not prisoners but still had the same needs as every other refugee. Thus the ad hoc nature developed; the natural migration was towards the centres of population and the railheads and routes around the country.

Over time Kitchener expressed several rather contradictory aims for the camps,[74] including protection for loyal burghers, imprisonment of families to force the Boer men into surrender, safeguarding all the families from black Africans, and preventing suffering from hunger.

The first camps were burgher camps set up in September 1900 at Eerste Fabrieken east of Pretoria, to bring off the veld and protect those loyal to the British as refugees from the war. Later the National Scout families (Boers recruited to support and fight for the British) also had to be removed from the veld for their own protection, and were initially contained in a separate part of an internment camp. This was to cause so much friction between the two different sets of families that separate camps had to be established and these families suffered in a similar way to Boer families, without favour.[75]

By December 1900 Kitchener had taken the next step by issuing orders to bring into camps all women, children and Africans in areas where the commandos were active in order to limit the endurance of the guerrillas. However the wholesale sweeping up of families did not occur until after the failed meeting Kitchener held with Botha to negotiate peace in February 1901. It was a cordial meeting but ultimately failed to find any common ground on which to negotiate a peace settlement. Kitchener to his credit tried to negotiate that the remaining families on the veld be spared, despite their husbands being on commando, if Botha would agree to stop recruiting families who were either neutral or loyal to the British. Botha did not agree, and Kitchener was left little option but to remove these families into the camps. He later wrote to Botha, 'I have no other course open to me, and am forced to take the very unpleasant and repugnant steps of bringing in the women and children.'[76] This suggests that much of the internment of families off the veld, especially relating to loyal burgher families, was not a part of a co-ordinated strategy, but forced upon Kitchener by his inability to reach a settlement with Botha.

Despite the confusing aims for the camps, their use did have some effect on the combat capabilities of the commandos and their morale as the war progressed through the guerrilla phase. For Roberts the camps started as temporary affairs, and so they were not considered worthy of planning or resourcing; this notion sadly was retained for some considerable time. Taking precious military staff away from the war effort to look after civilians was anathema to the British military of the period. As the issue of the camps grew beyond all conceivable proportions, Kitchener sought to offload the responsibility onto a civilian administration, and issued an order to this effect on 7 March 1901.[77] The military retained the responsibility for the camps' external protection, however, hence many blockhouses were situated close to camps.

Visitors to South Africa such as the famed Emily Hobhouse, intent on reviewing conditions and reporting back to parliament, only served to further highlight the soaring death figures. This was the most likely reason for a policy reversal by Kitchener, and in December 1901 he ordered the columns not to take any more civilian prisoners. Viewed as a gesture to the Liberals, on the eve of the new session of Parliament at Westminster, it was a shrewd political move. It also made excellent military sense, as it greatly handicapped the guerrillas, now that the drives were in full pursuit of the Boers.[78]

DENIAL OF MOVEMENT

So commenced the final phase of using blockhouses in June 1901, with an effort to form a mesh or net of blockhouse lines stretching across country and connected to the railway blockhouse backbone. The aim was to divide the open veld up into areas, or pens, into which the Boer commandos could be driven. The defensive nature of the system was now to be used in an offensive mode in which the Boers would be trapped between roving British columns and blockhouse lines. Construction entered its last frenetic and controversial phase, where lines were extended using the lie of the land by placing blockhouses within rifle range of each other and connecting them by trenches and wires. Towards the end of the war Kitchener was even advocating offering a bounty of £5 for anyone who caught a Boer using what he now termed the 'enclaves' formed by the blockhouse lines as catchment areas.[79]

The longest single line was in the Cape Colony, connecting Victoria West to Lambert's Bay via Carnarvon, Williston, Calvinia and Clanwilliam. Allowing for the terrain, this was over 350 miles (560 kilometres) and the author calculates it used approximately 500 blockhouses built from January 1902 to May and the end of the war. Today along this line only one replica at Carnarvon and a relocated original at Graafwater Farm remain. They were supposed to provide a barrier to protect the Cape from marauding commandos but posed only a thin line that was largely ineffective, stretched across rugged terrain.

Now that the blockhouse was an industrialised product, engineered blockhouse factories were established at Middelberg, Bloemfontein, Elandsfontein rail junction near Johannesburg, Standerton, and at the ports of Cape Town and Durban.[80] Engineering companies were slow to adopt the new designs and early versions varied greatly until small models were sent out to all companies and some uniformity established. Blockhouse kits were then sent out with instructions and sappers supervised military and black building parties. So mobile were these structures that whole lines could be collapsed and re-erected at another place. Often refurbishment took place and the blockhouses were re-kitted at the manufacturing depots if spares were needed for damaged parts.

THE DRIVES AND NEW MODEL DRIVES

The first of the early style drives introduced by Kitchener was undertaken on 29 November 1900. It was defined as a systematic drive organised like a country game shoot to flush out Boers, using the troops as beaters. Weekly success was defined by the number of Boers 'bagged' that week. The large roving columns used to scour the veld were also used to eradicate all support for the Boers and they cleared farms, families and black workers alike. Sometimes they acted on intelligence and were active in hunts for a particular commando, but often ended up as futile roving searches. The first few of these 'hunts' in February 1901 were aimed at capturing the notorious De Wet, and could be extremely complex, involving up to 17 columns.[81] When Kitchener took over command there were 38 of these large and cumbersome columns; they provided little in terms of result and the infamous De Wet remained at large.

Once the blockhouse lines were developed, with the railway-defended backbone and cross-country lines, the tactic of 'driving' favoured by Kitchener could be more effectively employed. The initial concept was now developed into a more systematic approach using the blockhouse lines as barriers against which to trap the commandos. It was similar to the tactic employed previously to try and capture the roving Boers, but now the concept had more structure. The British employed large roving columns, now deployed as a dragnet, soldiers almost shoulder to shoulder in an attempt to push the Boers into a corner.

The methodology of the 'New Model Drive', was to choose an area where Boers were bounded by a blockhoused rail line and a cross-country blockhouse line. The bounded area formed an enclosed perimeter in which to start driving at one end and in the case shown in the diagram drive them into the spout of a gradually decreasing funnel. The start line formed by the five columns marked A to E covered an initial frontage of perhaps 60 miles (96 kilometres). Each day the drive covered about 20 miles (32 kilometres) and then stopped and camped for the night, laying out a chain of outposts to secure the dragnet until it continued the next day.

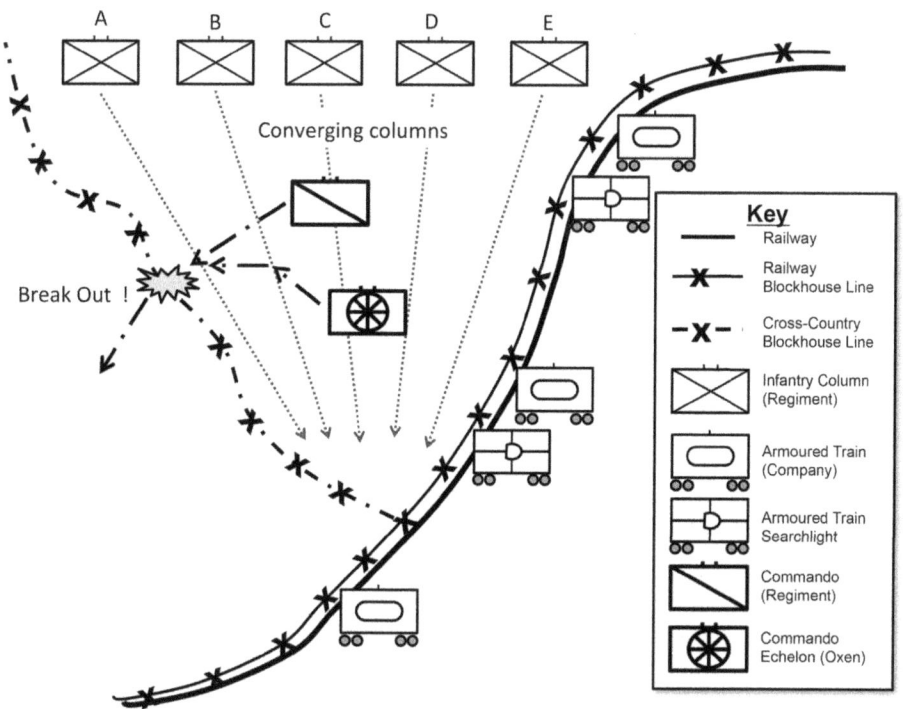

Schematic of a New Model Drive, and a Boer breakout.[82]

The new tactics also called for differently configured columns, 90 of them by the time the war ended. New Model Drive columns were much more intelligence led: the black scouts and Boer *hensoppers* who manned the National Scouts had tighter command and control exercised over them, and their movements were better coordinated. In addition, they stripped down for the task at hand, often only carrying one day's animal forage, two days rations, and 150 rounds of ammunition. The aim was to get to a known Boer laager quickly and be ready to surprise them at first light.

The first of these New Model Drives took place rather late in the war, on 5–8 February 1902. It involved a considerable force of about 17 000 troops, 300 blockhouses, and 7 armoured trains. The stage was set to capture De Wet's commando,[83] which was still a thorn in the Army's side. His victory over the British at the Battle of Groenkop in December 1901 had caused a real blow to Kitchener who then reinvigorated the building of blockhouses in an attempt to hamper movement even more. He saw De Wet's capture as a potential 'war winning' activity, which would certainly bring an end to hostilities at the very least in the Orange Free State.[84]

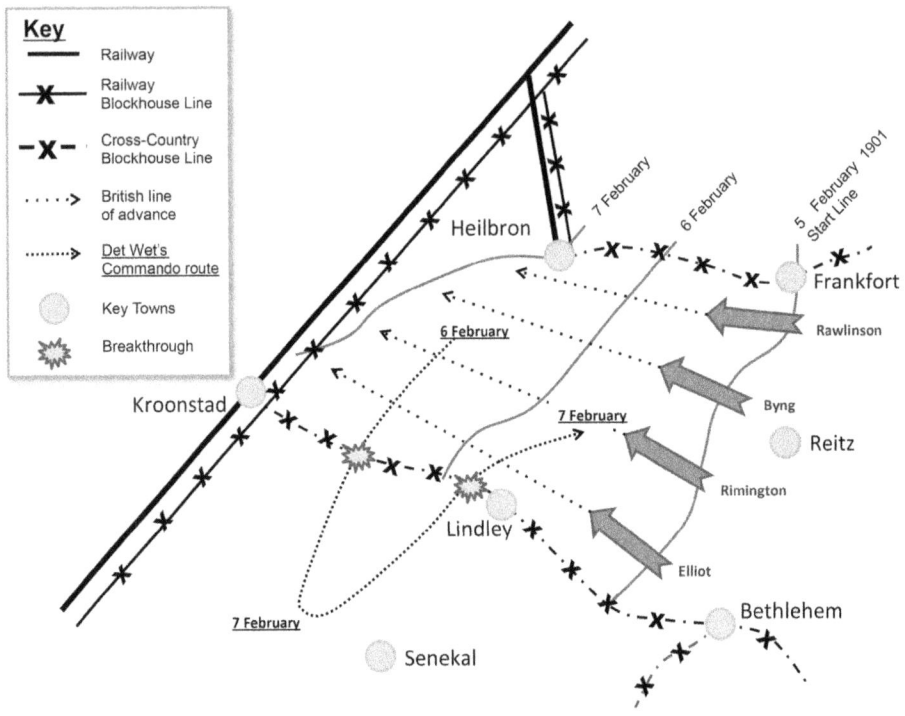

The first drive against De Wet, 5–8 February 1902.[85]

The Boers, however, were able to counter this concept in two ways. Either they would wait until darkness and then try to slip through or rush the guards; or attempt to break through the blockhouse lines at a weak point. De Wet had selected a point on a cross-country blockhouse line between Lindley and Kroonstad, and passed the entire commando through the line without a shot being fired. Several of his men actually failed to realise that they had passed the British lines and were still nervously waiting for a fight! In the same crossing the rear echelon of the commando brought a herd of cattle through which stampeded and caused some engagements from the blockhouses. This incident led to the newspaper report that the Boers were using cattle to forge a passage through the fortified lines. In fact this was not true, as De Wet explained in his post-war memoir: 'This paper declared that I had driven a great herd of cattle in front of me to break down the fencing! ... This is the way the English write reports.'[86] The cattle were too precious to waste doing this, and the lines quite easy to penetrate, but it made for a good story. For the size of this first drive and immense British effort, 286 Boers were killed, wounded or captured.[87]

A German postcard of the period immortalising the supposed use of cattle to break the blockhouse lines.[88]

The ease with which De Wet passed through the blockhouse line was commented on by Ian Hamilton, who gives some idea of the weaknesses of this concept: 'He escaped through the Lindley blockhouse line with about 50 men, halfway between Lindley and Kroonstad. The wire was found cut within very few yards of a post belonging to a regiment specifically interpolated between two blockhouses, at that spot to strengthen it. The men all stoutly swore that they had neither heard nor seen anything, so I fear the only inference is that, in spite of the most earnest warnings to keep alert that night, they were sound asleep, or else and this is more likely, they ducked the Boers and did not dare to fire.'[89] The only saving grace was that the regiment was not specifically named and shamed. The lack of moral fibre to engage the Boers as orders demanded was an issue for the chain-of-command for both Imperial and African components.

Although the effort-to-success ratio in terms of men seemed disproportionate, Kitchener considered this an effective tactic and a successful operation. He immediately ordered another larger drive to take place a few days later, from 13 to 27 February 1902. Again De Wet's commando was the target, and this time Kitchener's force managed to engage with the Boers. Although De Wet managed to evade capture once again, the 'bag' was much higher, with 828 Boers killed or captured, and 25,000 head of cattle and 200 carts and wagons captured. This had been more of a rout, and highlighted the enormous logistical tail that was vulnerable whilst 'on commando'.

The third drive again came quickly. Kitchener, once seized with a method of success, sought to exploit it ruthlessly. Although the men were exhausted from marching day after day, between 5 and12 March 1902 the columns went back over the first area again, 'bagging' a further 100 Boers.

Previously columns had been allowed to do their own thing; now there was a need for an intermediate level of command, and Ian Hamilton was sent from Kitchener's staff in April 1902 to take up this duty. He was able to utilise intelligence in order to have multiple columns searching, changing their direction according to the terrain and the updated intelligence picture on Boer movements.[90]

This method was really a sledgehammer to crack a nut, but Kitchener had the forces in sufficient size to make this tactic work. It was a war of attrition; gradually it was reducing the combat effectiveness of the commandos, cutting their numbers and sapping their integral supporting echelons. The Boers were wearied by the tactic, but ultimately did not really fear it, as the blockhouse lines and night pickets were easily breached. No Boer leader was ever captured during the drives; they proved to be too agile.

The lines were long, thin, easy to spot in the open veld and easy to rush at night and pass through. There were some successes when commandos were cornered, but these captures were in a minority to the Boer forces. The blockhouses played their part in the counter-mobility operation, but were also relatively easy to pass through, though some Boer leaders found it less easy than De Wet. In terms of effectiveness, the blockhouse lines and drives harried the commandos, cramped their initiative and mobility and over time wore them down materially and reduced their morale.[91]

New Model Drives against De Wet	Boer casualties & POWs
Drive 1 5-8 February 1902	286
Drive 2 13-27 February 1902	828
Drive 3 5-12 March 1902	100
Total	1,214

Summary of the first New Model Drive 'bags'.

COUNTER-MOBILITY OPERATIONS

A light, mobile commando, agile and familiar with terrain, could pass through a fixed blockhouse line or evade a large and slow-moving 'driving' column strung out across the veld. Even larger commandos with small carts and ox herds had also passed through these British lines. What the Boers feared most, however, was equally equipped light and mobile forces. At the height of the war the British deployed over 80,000 troops in a mounted infantry role; supplied with ever-improving intelligence they could be far more effective at targeting a commando than the blanket sweeps being employed on the New Model Drives. This was what generals such as De Wet feared the most.

In a Top Secret telegram to Lord Roberts in November 1901 Kitchener muses: 'The blockhouse system does a great deal but takes up a large number of infantry. I should like to extend the line considerably also to deal effectually with the most difficult question of how to make troops more mobile. I have tried to pack saddles and have to an extent succeeded but it is not enough to catch the Boers who go with nothing and trust to pick up a few mealies in a native hut.'[92]

Converging Mobile Columns

Evidence from the writing of De Wet gave the view that the commandos feared the use of converging columns much more than the well-documented and orchestrated drives against blockhouse lines. Lighter and more mobile columns when deployed in concert to target a single commando sought to engage the Boers using counter-mobility as their key doctrine. The blockhouse lines were thus somewhat ineffective as they were static, their weaknesses could be reconnoitred and lines breached at night with relative ease. The continuously moving columns however were less easy to fix and avoid, especially when there were several of them in different directions. De Wet in his memoirs recounts: 'The English have been constantly boasting in their newspapers about the advantages of their blockhouses, but have never been able to give an instance of a capture affected by them. On the contrary when during the last stages of the war it happened, as it often did, that they drove some of our men against one or other of the great blockhouse lines which then intersected the country, and it became necessary for us to fight our way through, we generally succeeded in doing so. And that, with fewer casualties than when, as in the instance I have just given, they concentrated their forces, and formed a circle around us.'[93]

In terms of effort expended compared to the results obtained, never did the two seem balanced or in favour of the British. One such example was that of Colonel Rawlinson's column,[94] which comprised two battalions of mounted infantry, and for part of the time the Imperial Light Horse. In 14 months from April 1901, it covered 5,211 miles (8,390 kilometres) and made 276 camps, for a return of 151 Boer casualties, 1,376 POWs and the capture of 3 guns, 1,082 rifles and 68,600 rounds of ammunition, for only a loss of 12 British deaths and 42 wounded. This indicates the level of effort to cover the ground and how long it took to gradually wear down the commandos. Other columns' statistics were quite similar and also noted the quantities of livestock captured and destroyed. Capturing 5,000 cattle and 15,000 sheep was typical, and then having to destroy them because the driving was too difficult was not uncommon, which further frustrated the commando's ability to sustain itself on the move.

One of the more successful columns and most feared by the Boers was one led by Lieutenant Colonel George Elliot Benson.[95] He was a Royal Artillery officer, who had passed staff college and had many years of experience in the Sudan and Gold Coast, and had been in South Africa virtually from the beginning. His success was largely down to the use of intelligence to steer his column into the most effective areas to capture Boers. He had an effective intelligence officer, whom he was prepared to resource and trust, making the greatest use of his knowledge and advice. The Intelligence Officer was Colonel Woolls-Sampson, and he had recruited a small force of black scouts and spies, which he used as a screen for the main force and a reconnaissance capability to locate Boer laagers to good effect. Out of the 28 attacks mounted by Benson, 21 were successful in their aim of destroying or capturing commandos. The column was a fast-riding, hard-hitting unit, quick to take the lead from good intelligence to catch resting Boers largely unprotected. It was 1,400 troop strong, with four guns and two pom-poms, which were eventually defeated by a large coordinated attack of several commandos in October 1901.

Night Attacks

Scouts and spies, including some Boer collaborators, provided a steady flow of intelligence, allowing for a more focused and incisive use of force to attack the commandos. Often the attacks came in the early morning, before sunrise, catching the sleeping Boers in their laagers. Later in the war the British employed the National Scouts (Boers sympathetic and recruited to serve in the British Army as scouts) who knew the vulnerabilities of the night-time laager. By De Wet's own admission, 'at last they had found a way of inflicting severe losses upon us'.[96] This is more evidence of the resolve to adapt and change strategy and tactics and fight the determined and elusive Boers, a term coined for modern warfare of nascent doctrine or 'doctrine on the fly'.

> **What is Nascent Doctrine?**
>
> **Definition of Nascent**: Nascent (A); Emerging; just coming into existence, young.[97]
>
> **Definition of Doctrine**: Doctrine (N); A belief or tenet, especially about philosophical or theological matters.[98]
>
> **Definition of on-the-fly**: On-the-fly (A): something created when needed; something that was not planned ahead; or changes that are made during the execution of some activity.[99]
>
> In defining 'doctrine' there are many references to religious sources, which would certainly please those 'Doctrinal Gurus' who serve in every Army across the world. To move from the purest definition to that more focused to the military application, military doctrine might be defined as 'the concise expression of how military forces contribute to campaigns, major operations, battles and engagements'.[100]

ARMOURED TRAINS

Soon after Lord Kitchener assumed command he decided to organise the armoured trains as fighting units, and appointed to his staff an officer titled the Assistant Director of Railways for Armoured Trains. This officer was also on the staff of the Director of Railways, and was placed in charge of all the armoured trains in South Africa, some 20 in number. The officer concerned was Percy EPC Girouard, a French Canadian by birth, who had achieved some prestige in the Sudan Campaign. There, as a young lieutenant, he had worked for Kitchener to make railways a key strategic component of the desert campaign.

Initially trains were used to good effect as the key line of supply as Roberts's Army moved to capture Pretoria; thereafter Kitchener deployed 20 armoured trains to support operations during the guerrilla phase. The principal duties[101] of these armoured trains were to:

1. Act in conjunction with columns in the field, to intercept the enemy whom the columns were driving on to the line.
2. Act on the flank of a column or line of columns, the train being well advanced so as to prevent the enemy breaking to that flank.
3. Reinforce stations and camps on the railway, which were threatened by the enemy.
4. Escort ordinary traffic trains.
5. Conduct reconnaissance.
6. Patrol by day and night.
7. Protect general traffic routes.

Interestingly, there is no mention of a duty to provide direct support to blockhouses; however points 1 and 2 clearly indicate the obligation. Armoured trains were always vulnerable to attack and required some sort of 'eyes and ears' in front of the train to forewarn of the enemy. Similarly, when working along a railway blockhouse line, rockets were the key means of signalling to the train for assistance. One rocket was 'enemy attacking'; two rockets was a call for help. Trains were encouraged to move as much as possible during the night, but were required to halt at pre-arranged times to connect and communicate on the telegraph system. This was essential and to be done several times a night so as to get up-to-date intelligence and receive new orders. The garrison of an armoured train was what today would be called an 'all-Arms' unit made from a number of constituent parts. In addition to the infantry escort, it contained Royal Artillery and Royal Engineers detachments. The latter consisted of one non-commissioned officer and six sappers or privates skilled in railway repair work and in re-setting derailed engines and trucks. There were also two telegraph linesmen, one telegraph clerk, two engine drivers and two firemen. All the men of this detachment were counted as infantrymen when the train was engaged, with the exception of the driver and fireman on the footplate. Even the latter carried rifles in the engine cab to drive off an enemy trying to gain possession of their engine. Additionally they had a powerful telescope, useful for daytime observation. The train if halted on a prominent rise might command extensive views across the veld. Trains had mounted scouts who were used from halts at intermediate sidings to reconnoitre; some also had 'native scouts.' Horses were accommodated in the cattle truck on the train. This enabled an early response to threats and the ability to see into dead-ground and reverse slopes where enemy may be hiding.

Eduard Percy Cranwill Girouard[102]

Sir Percy Girouard K.C.M.G.[103]

Eduard Percy Cranwill Girouard was born on 26 January 1867 in Montreal, Canada, being commissioned in the Royal Engineers in 1888. His family wanted him to remain in Canada to run the railways there, but he had a 'wanderlust' to build railroads throughout the British Empire. On commissioning he had run the Royal Arsenal railway from 1890 to 1895, to great effect and reputation. Percy Girouard first came to prominence during the Egypt Campaign of the 1890s, when he rose to prominence as a young engineering officer. This must have been quite a sight, as he worked for the somewhat dour and forthright Kitchener, who had already recognised his ability and reputation as a railroad builder. Kitchener met Girouard in London prior to his move to Egypt and arranged for his transfer to the Egyptian Army and railway in 1896. Both of course had engineering roots. Girouard was a high-spirited, handsome and cheerful young man who was known to speak his mind, and was known by Kitchener as 'Gerry'. In Sudan[104] many railway and military experts were sceptical of the use of railways in the theatre of operations, considering them rather impractical. Kitchener however put his trust in Girouard, who made railways a vital part of the push through the desert. The Sudanese Military Railway from Wadi Halfa to Abu Hamed was built across the Nubian Desert, some 217 miles (350km) of narrow gauge line of 3ft 6in (1.06 metres). It allowed Kitchener to bring his Anglo-Egyptian Army of 23,000 men into the heart of the Sudan, and ultimately the relief of Khartoum, and sealed his fame and success. Victorian writers at the time described it as the greatest military weapon ever forged against Mahdism. During the campaign Girouard was twice mentioned in despatches for 'efficient performance' and then for 'completion of the line in record time, under great vicissitudes and during extremely hot weather'. For this campaign he was awarded a Distinguished Service Order (DSO) in 1896 and later knighted with a KCMG (The Most Distinguished Order of St Michael and St George) in 1901. It was no surprise therefore that 'Gerry' would next deploy like many of his counterparts into the South African theatre. In 1899 he became Director of Imperial Military Railways working under Roberts and Kitchener. Girouard, now a captain, and knighted, served in a similar role managing and directing railway infrastructure during the Anglo-Boer War. Initially taken aback by the sheer scale of infrastructure destruction by the Boers, he soon set about bridge repairs but could hardly keep pace with the rate required for the advancing Army. In making this infrastructure available, it allowed Roberts to secretly use rail transport and detrain his troop between the Orange and Modder Rivers. This was over 30,000 men with horses, mules, oxen, guns and transport[105] and the largest troop move of the campaign. Girouard claimed: 'This railing of troops was naturally a great advantage to us over the enemy.' For this work he was again mentioned in despatches.

Later he was instrumental in securing the vital supply lines north during the British advance. The rail line north at this point was vital in keeping the Army on the move and was a key target for Boer sabotage. At one point the line between Bloemfontein and Kroonstad was blown up 17 times,[106] a significant threat, as Roberts's Army needed two train loads of supplies per day to keep it on the move. It was Girouard who again came to the rescue and under his direction the lines were repaired and operations continued. For this he was twice more mentioned in despatches.

He completed the war as a lieutenant-colonel and took charge of reconstructing the railways of both Republics until 1904, when he resigned due to Afrikaner hostility to him. He continued his military career largely in railways working for the Colonial Office in Nigeria, Kenya and elsewhere in East Africa. He then became Director General of Munitions Supply in 1915, but resigned in 1917 to join the Board of Directors for Armstrong-Vickers. He resigned from public life in 1919 and died on 26 September 1932 in London.

The Train Officer Commanding had a formidable weapon system at his disposal, capable of attacking superior forces, but also had much vulnerability. Trains had a variety of armaments consisting of one large field gun, usually a 12-pounder, two Maxim machine guns, a detachment of infantrymen, of section strength, and later in the war a large searchlight was added. The field gun provided significant artillery firepower, while the two machine guns (one at either end of the train), arranged with lateral sweeps, allowed the fire support to cross each other 50–80 yards (45–70 metres) either side of the train. These two machine guns could then provide protection to the flanks of the train, both on the move and static as the operation dictated.

There was a constant danger of the line being cut from contact or remotely detonated mines. Often the lead truck would take the blast, which would be followed up quickly with an assault on the wrecked train and dazed occupants. To counter this each train was ordered to push a heavily laden truck in front of the engine, and doubled up to carry spare sleepers, rails and telegraph poles.

One such attack was against No 6 Armoured Train, which failed to be pushing its mine-counter truck and exploded a mine near Kroonstad. The Officer Commanding was killed instantly, the leading fighting truck was overturned, and several men in it were injured. A few days later this same train, having again been put back in commission, ran over a contact mine near Heilbron. On this occasion the front truck was in place and fired the mine. Only a length of three feet (1 metre) of rail was blown out, but as the mine was laid on a straight portion of the line, the whole train bumped across the break and kept the rails. Three minutes after the explosion it was engaging the enemy with the 12-pounder quick-firing gun. There were no casualties on the train during this attack.

No 18 Armoured train base in the America area.[107]

Armoured trains were also moving telegraph offices, equipped with field sounders, vibrators, phonophores and telephones, all considered modern technology of the day. When trains stopped away from a regular static office at their siding or other tactical location, which they did nearly every night, they were never out of communication with the neighbouring stations or blockhouse lines. When several trains patrolled a particular section, it was found advisable, especially at night, that they should all halt at fixed intervals and connect up with the telegraph wires to receive instructions and news. Regular communications on the move during this war were not yet possible, despite the advent of wireless communications during the conflict.

Armoured train and blockhouse at work in a stylised depiction of De Wet's crossing the railway.

One of the improvements made to armoured trains later in the campaign was the addition of a strong electric searchlight. A total of 16 of 20 trains were fitted, with four of the trains having two searchlight projectors. Initially the lights were placed above the leading truck, moving from one end to the other at a siding when the patrol direction changed. In the centre it proved ineffective and difficult to control. The steam from the train's engine and turbines worked the dynamos to produce electricity for its operation. These were quite effective and disrupted many a Boer crossing, as recounted here by General Ben Viljoen in his own words:[108]

We approached the line between Balmoral and Brugspruit, coming as close to it as possible with regard to safety, and we stopped in a 'dunk' (hollow place) intending to remain there until dusk before attempting to cross. The blockhouses were only 1,000 yards distant from each other, and in order to take our wagons across there was but one thing to be done, namely, to storm two blockhouses, overpower their garrison, and take our convoy across between these two. Fortunately there were no obstacles here in the shape of embankments or excavations, the line being level with the veld. We moved on in the evening (27 June), the moon shining brightly, which was very unfortunate for us, as the enemy would see us and hear us long before we came within range... I had arranged that Commandant Groenewald was to storm the blockhouse on the right, and Commandant W Viljoen that to the left, each with 75 men. We halted about 1 000 paces from the line. When 150 yards from the blockhouses the garrison opened fire on our men, and a hail of Lee-Metford bullets spread over a distance of about 4 miles – the British soldiers firing from within the blockhouses and from behind the mounds of earth. The blockhouse attacked by Commandant Viljoen offered the most determined resistance for about 20 minutes, but our men thrust their rifles through the loopholes of the blockhouses and fired within.

The first blockhouse surrendered with the loss of one of Viljoen's men but the second blockhouse was built of stone and offered much greater resistance. Viljoen continues with his account:

Many of our men had fallen and an armoured train with a searchlight was approaching from Brugspruit. On the other side of the blockhouse we found a ditch about 3 feet deep and 2 feet wide. Hastily filling this up, we let the carts go over. As the fifth one had got across and the sixth one was standing on the lines, the armoured train came dashing at full speed in our midst. We had no dynamite to blow up the line, and although we fired on the train, it steamed right up to where we were crossing, smashing a team of mules and splitting us up into two sections. Turning the searchlight on us, the enemy opened fire on us with rifles, maxims and guns, firing grapeshot. Commandant Groenewald had to retire from the unconquered blockhouse, and managed somehow to get through. The majority of the burghers had already crossed and fled, whilst the remainder hurried back with a pom-pom and the other carts. I did not expect that the train would come so close to us and was seated on my horse close to the surrendered blockhouse when it pulled up abruptly not four paces from me. The searchlight made the surroundings as light as day and revealed the strange spectacle of the burghers, on foot and on horseback, fleeing in all directions and accompanied by cattle and wagons, whilst many dead lay on the veld. However, we saved everything with the exception of a wagon and two carts, one of which unfortunately was my own.... The 'Tommies' fired furiously from the blockhouses and our friend the armoured train was seen approaching from Middelburg, whistling a friendly warning to us. It came full speed as before, but only got to the spot where the mine had been laid for it. There was a loud explosion, something went up in the air and then the shrill whistle stopped and all was silent.

This account gives a good description of how the Boer commandos were able to exploit the static vulnerability of the blockhouses, and their lack of ability to support each other with interlocking fire when directly attacked. If the Boers had the benefit of surprise and the operation was quickly executed, blockhouses could be overpowered and ditches filled in to overcome the barrier. However if the attack became protracted for any reason the Boers were then vulnerable to support in the form of the formidable firepower from the armoured trains. In this case the train was deployed aggressively into the midst of the Boer force!

In another example armoured trains fitted with searchlights were used effectively on The Drives, working ahead of the columns. In one drive during early 1902, four searchlights were placed at intervals on the Natal Line to work in conjunction with columns driving onto the railway blockhouse line. Although these lights effectively turned the night into day, Royal Engineer reports tell of railway blockhouse line crossings being made within 5 miles (8 kilometres) of the light, in a hollow in the dead-ground which the light could not reach.[109]

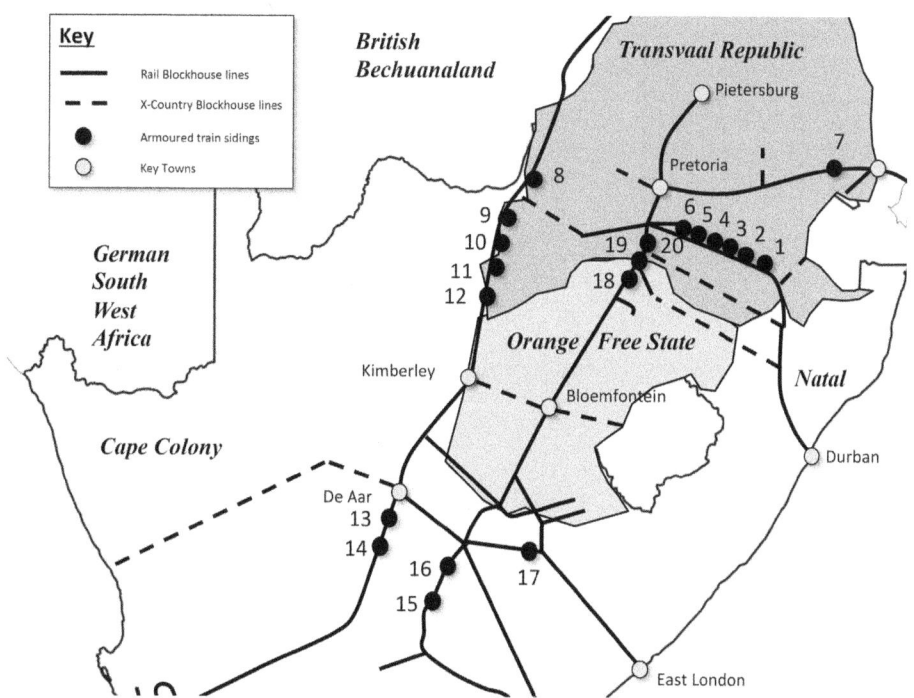

The locations of 20 armoured trains at the end of March 1902.[110]

THE BIRTH OF COUNTERINSURGENCY

The asymmetry of the opposing forces and the nature of the Boers' resolve to fight the British Empire really led to the birth of what we understand today as Counterinsurgency Warfare. It also highlights many of the lessons required to effectively fight a guerrilla-style enemy; sadly for the Empire of the day, it was still organised and mentally configured to fight large-scale or colonial wars. The vastness of the terrain and the forces deployed to counter what was effectively a roving force of around 40,000 Boers was considerable. The British and Colonial Forces troop turnover in-theatre was up to 450,000, outnumbering the enemy by ten to one. The British lost 7,792 soldiers as a result of battle with a further 14,658 deaths from disease or accidents, and 73,430 were sent home ill or wounded.[111] The Boer losses in battle were 3,990, with an additional 157 deaths from accidents and a further 924 from battle wounds or disease, with a further 1,118 deaths as prisoners of war, making a comparable total of 6,189 in stark contrast to British losses. At a cost of over £1.5 million per week[112] the war ground on to an eventual capitulation of the Boers, and a British victory, but at a considerable price paid in lives and integrity.

There was also a significant cost to the black participants. Both the opposing forces might be classed as colonial invaders, and local impact on the black community has taken time to appreciate and realise. The Centenary memorialisation led by a government now comprising a black majority sought to correct this and museums and memorials show more reverence to those who fought and died for each side and those who also died in black concentration camps. The outcome of the Second Anglo-Boer War, and the defeat of the Boers, also laid the foundation of what came later throughout the two World Wars and into post-World War South Africa and the Nationalism that followed.

VICTORY IN THE END

The so-called 'last of the Gentleman's Wars' and 'white man's war' turned out to be euphemisms for days gone by, coined at the start of the conflict and hardly appropriate today and considering the outcome. A war on the cusp of a new century and over a century later is best remembered for being unique for its time, and also for many issues remarkably poignant for today's military strategists in contemporary wars such as Iraq and Afghanistan.

The war had to end; it was politically unpopular, there had been staggering defeats, instances of military incompetence and a parliamentary outcry at the treatment in the concentration camps. Also there was gold and substantial mineral wealth to be had, and the joining of two white minorities was important if the black workers were to be subjugated as the workers to mine them. The Peace Treaty of Vereeniging was signed on 31 May 1902, and although not popular with all Boers, the agreement provided for the end of hostilities and eventual self-government for the Transvaal and the Orange Free State as colonies of the British Empire.

The Boer Republics agreed to the sovereignty of the British Crown and the British Government agreed to the clauses of the Treaty:[113]

1. All Boer fighters of both Republics had to give themselves up.
2. All combatants would be disarmed.
3. Everyone had to swear allegiance to the Crown.
4. No death penalties would be dealt out.
5. A general amnesty would apply.
6. The use of Dutch would be allowed in the schools and law courts.
7. To eventually give the Transvaal and the Orange Free State self-

government (civil government was granted in 1906 and 1907, respectively).
8. To avoid discussing the native enfranchisement issue until self-government had been given.
9. To pay the Boers £3,000,000 in reconstruction aid.
10. Property rights of Boers would be respected.
11. No land taxes would be introduced.
12. Registered private guns would be allowed.

Although there is no specific mention of them, over 8,000 blockhouses were redundant overnight and had to be disposed of back to the community. Chapter 10 discusses in detail what happened, and how 120 years later less than 100 of these blockhouses remain as monuments of a war long gone.

THREAT AND COUNTER-THREAT & DEVELOPMENT OF DOCTRINE

The requirement for British generals over 100 years ago to change and adapt their tactics and equipment rapidly against Boer guerrilla forces is not lost on today's modern war fighter. Warfare is a continually changing and fluid dance of threat and counter threat, and never more so than with guerrilla opponents. To counter the threat of the improvised roadside bombs in Afghanistan, from a similar asymmetric enemy, the US military had to procure and deploy 12,000 Mine Resistant Ambush Protected[114] (MRAPs) vehicles in 18 months. Procurement was initially undertaken largely 'off-the-shelf' from private companies and most certainly as an urgent operation requirement, leaving training and doctrine of how they were to be operated to play catch-up, in theatre.

Rice Pattern blockhouses similarly were procured quickly and were certainly not part of the overall initial operational budget. Noted for their simplicity of design and build, the materials also were largely unavailable in theatre in the quantities required. Much of the corrugated iron and wire was sent from the home base in Great Britain, and both military labour and local contractors used to rapidly deploy the network. Although simple in concept to 'operate', with a low threshold of training required, their conceptual and doctrinal impacts can only have grown as their deployment extended, from a point defence to a long thin defensive system. There was also a negative morale component that was underestimated in placing largely active and mobile troops on static defensive duties for long periods of time. The moral of the day was to just to endure with a stiff upper lip.

New requirements such as the MRAPs, specific to operational needs at the time, are supposed to be decommissioned after the event, as were the Rice Pattern blockhouses; but more often than not these days expensive technology has to be brought into the military inventory and has to be sustained. Post-war activities saw the rapid decommissioning and selling off of blockhouses, with no contemplation of re-use; their day was done; however the MRAPs carry on protecting lives long their first year, when over 300 roadside explosive attacks failed to take one life.

The similarity is to get to know your enemy, his weaponry and tactics quickly and then adapt equally quickly in order to counter them with your own new equipment and doctrine. This rapid adaptation is often referred to as doctrine 'on-the-fly' or nascent doctrine. Nick Clegg MP, in a speech to Chatham House,[115] also refers to the intervention in Kosovo in 2001 as 'nascent doctrine', defining the sanctions and military measures as preclusive intervention against a regime that had 'radically compromised its Sovereignty by its own acts'.

Roberts left the theatre in December 1900, clearly believing the then logical doctrine that if a nation's capitals are seized then the war is won. Kitchener was to be left with 'mere mopping-up' operations for the remaining scattered commandos. The Boers had other ideas; the British were not fighting a native campaign, but one against dedicated and determined men and women who had in many cases fought for the right to be there in their own colonial war with the Zulus and other tribes. Thus, they embarked on a guerrilla war, and the British had to rapidly develop their counter-strategy and tactics in order to combat this strategy.

Having trained in the intervening years between wars, armies are often better able and ready to fight the next one – but only if it resembles the previous one. Now the British parade ground army had to adapt to 'new' Boer tactics and do it quickly. The large garrisons and columns could easily be out-manoeuvred by small mobile forces; in meeting engagements they could be out-shot through better marksmanship; Boer local intelligence and knowledge of the ground was also much better. Scorched earth, concentration camps, the blockhouse system and the drives were the key counters used by the British.

The generals – and Kitchener in particular – had to take action based on previous knowledge and experience and work it out 'on-the-fly'. The Second Anglo-Boer War was an important war in this regard, and led to further Army reforms, which were urgently needed. Was it ready for the next challenge in the First World War? This is another story with topical relevance as the world remembered the 100 years since the commencement of this epic global struggle.

CHAPTER 4

EVOLUTION OF THE BLOCKHOUSE SYSTEM IN SOUTH AFRICA

Blockhouse Street is a street to remember in your prayers, a deadly, soul-destroying, damnably dull street of galvanised iron prisons, in each of which are six prisoners waiting for execution.

EDGAR WALLACE, *DAILY MAIL* REPORTER[116]

Throughout the British involvement in South Africa prior to the Second Anglo-Boer War, fortifications and blockhouses had played an important part of their defensive plans. In 1795, after the invasion of the Cape, Major-General Sir James Craig ordered three blockhouses to be built on Table Mountain to strengthen Cape Town's defences. Queen's Blockhouse (formerly the Duke of York's Blockhouse, then renamed after King George III's Queen Sophia Charlotte), Prince of Wales Blockhouse and the King's Blockhouse were built high on the slopes of Devil's Peak to defend the rear of the port.

The King's Blockhouse was named after George III of Great Britain and located on a strategic point to see False Bay to the southeast as a likely approach to Cape Town. From this position it was possible to send signals via the other blockhouses to warn the Castle of Good Hope of an impending attack. It is the best preserved of the three, and sits high on Mowbray Ridge beneath the steep rock bands of Devil's Peak. At the beginning of the 20th century the stone tower was used to house convicts who worked on the reforestation of the mountain.[117] Just below the blockhouse the British placed a Swedish 25-pounder Stavsjö gun dated 1782, a type of gun that would command an enormous range in such an elevated position, covering the seaward approaches.

The ruins of the Queen's Blockhouse are situated at the same level as the lower cable station, close to a road that traverses the mountain at a lower level than that of the King's Blockhouse, and the 25-pound Stavsjö gun now lies discarded without a carriage on the barbette wall. The last of the three blockhouses, the Prince of Wales Blockhouse, was named after George III's son who later became George IV, and today little remains other than the plateau and some of the barbette wall. It is situated much lower on Devil's Peak to the south of the other two blockhouses.

Another two towers were constructed during 1795: one in Simon's Town, to a similar design to that of the Martello tower, but with vertical and not sloping walls; and one named Craig's Tower (named after Major General Craig), which was demolished over 100 years ago. It was built to protect an existing seaward-facing gun battery on Salt River.[118] The Simon's Town Martello tower was restored in the 1970s for use as a naval museum, which has since closed to the general public and can only be viewed by appointment.

Also during the same period East and West Forts at Hout Bay were improved through the addition of a three-storeyed masonry blockhouse, which remains in a ruined condition today. They were originally built in 1783 by the French, for the Dutch, with soldiers of the French/Indian Pondicherry (Mercenary) Regiment, and housed eight cannons. In Cape Town itself the British took over the Castle of Good Hope, built between 1666 and 1679 by the Dutch East India Company as a maritime replenishment station. It was here between 1871 and 1873 that a young Elliot Wood as Commander Royal Engineers Western Province was billeted for a period;[119] perhaps his blockhouse ideas originated from studying the castle ramparts? All these defences in the Cape Town area however were largely to defeat an enemy who was most likely to be European and well armed with artillery approaching from the sea. Up to this point there had been no Great Trek north by the Boers, or British conflicts in Natal against 'native' armies.

FORTS IN THE FRONTIER WARS

Further up the eastern coast of South Africa the British later fought the Xhosa Wars (also known as the Cape Frontier Wars or 'Africa's 100 Years War'). They were a series of nine wars or flare-ups from 1779 to 1879 between the Xhosa tribes and European settlers in what is now the Eastern Cape. These events were the longest running military action in African colonial history. The period was marked by a progressive build and improvement of fortifications including blockhouses and signal towers throughout the region.

Typical of the many forts constructed during these wars was Fort Peddie This fort was built in 1835 on the banks of the Hlosi River and named after Colonel John Peddie, commanding officer of the 72nd Regiment (Seaforth Highlanders). In May 1846, during the War of the Axe, Fort Peddie successfully withstood an attack from about 10,000 Xhosa warriors led by Chief Dilima. Incredibly, not one of the people manning the fort was hurt.

Fort Peddie consists of a tower and cavalry barracks, both of which are still visible. The tower is one of many built to try and improve communication between Fort Beaufort and Fort Selwyn (Grahamstown). The idea was to build a line of towers within telescopic sight to communicate messages between each other. There were two lines known as the Peddie Line and the Beaufort Line. Communication was by semaphore, a type of signalling mast first developed by Claude Chappe during the French Revolution. The remains of a semaphore can be viewed at the top of the tower, though visitors should take their own ladder!

Fort Peddie Frontier War fort. Courtesy of Steve Moseley, Karoo Images.[120]

Several forts were constructed to this design throughout the region and many are in a reasonable 'state of survival' compared to the remaining Anglo-Boer War blockhouses, at 67 years their senior. They were generally rectangular in shape, their perimeter varying between about 380m (1247 ft) and 230m (755 ft), with loopholed masonry walls some 3 metres high, flanked by two towers at opposite corners and surrounded by a ditch.[121] They were constructed as places of refuge for local settlers, their families, wagons, and servants, and as safe havens, for those who needed refuge. Many are in remote locations and have certainly not attracted the interest of building material thieves or vandals, unlike the Anglo-Boer War blockhouses.

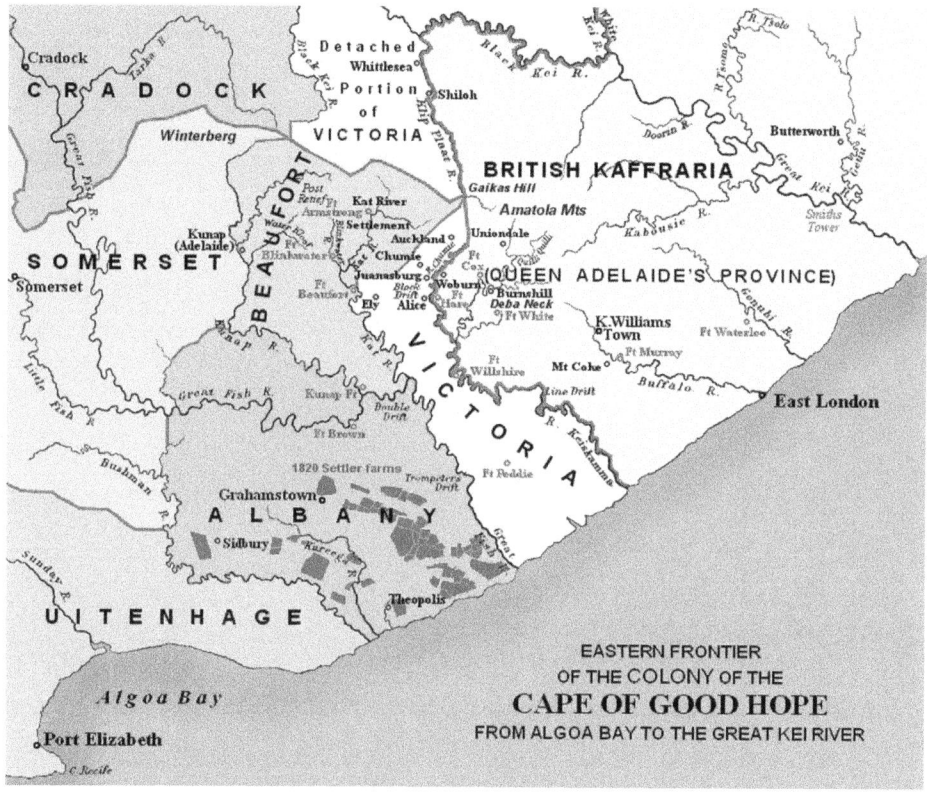

Frontier War forts on the eastern frontier of the Cape Colony.[122]

ANGLO-ZULU WAR 1879

In 1879 the British forces went to war against the Zulu kingdom, from the British-controlled area of Natal. Although the war was of British making and its army was outnumbered by a Zulu force of up to 40,000 men, Lord Chelmsford's force of 5,000 regular infantry and 20 field guns expected the campaign to be short. Even prior to conflicts with the Boer commandos, the British were to face a mobile and highly motivated force of Zulu regiments who knew the terrain and how to use it to their best advantage. The British were formed in large cumbersome columns with large supply trains and vulnerable supply lines, up to this point still dressed in their bright red dress tunics.

As the Royal Engineers had demonstrated previously they were more than competent at building well-designed and constructed stone forts and blockhouses. This campaign however demanded a different type of fortification in the form of fieldworks, which are defined as 'A temporary work constructed by an army in the field, used to cover an attack on a fortification, or as protection against another enemy army, especially a relieving force.'[123] Typically this was a temporary or tactical structure constructed from earth ramparts, wooden walls or stockades and dry-stone walls to protect both military garrisons and deployed troops. In addition they were also used as supply stations required to keep Lord Chelmsford's army in the field.

Army Standing Orders of the day stated that units were not to be broken down below company level into smaller detachments, and that fortifications required a rifle perimeter density of one man per yard (metre) approximately. This led to the construction of large fieldworks, often of two-company strength. A total of 20 were completed during the campaign of which 14 have survived urbanisation and farming activities, the most notable of which are the Eshowe Mission Station; and Forts Tenedos and Pearson on the escarpments overlooking the Tugela River.[124]

FIRST ANGLO-BOER WAR 1881

In the intervening years between the end of the Anglo-Zulu War and the short First Anglo-Boer War there seems to have been little fortification activity, however the British fortified Pretoria after they had annexed the Transvaal during the First Anglo-Boer War. After unsuccessful attempts to reverse the annexation by negotiations, the Boers started to lay siege to towns and Pretoria was fortified to repel these attacks. Three forts were constructed, Forts Royal, Tullichewan, and Commeline, in addition to fortifying the military camp, prison and the Loreto Convent which acted as the British headquarters.[125] A blockhouse is also recorded as having been built at Eloff Cutting north of Pretoria. The town did see some action, most notably for the winners of two Victoria Crosses, Lance-Corporal James Murray and Trooper John Danaher, who advanced for 500 yards under heavy fire from a party of about 60 Boers and brought out of action a private who had been severely wounded[126] at the Battle of Elandsfontein on 16 January 1881.

The Boers did however start a programme of militarisation after the First Boer War in 1881 and then after the failed Jameson Raid in 1895, in order to better defend their Republics. They built extensive fortifications to protect Pretoria and Johannesburg and equipped their military with modern artillery procured from Germany. The construction of four forts out of the original requirement of eight began in 1896 until 1898 for Forts Schanskop, Wonderboompoort, Klapperkop and Daspoortrand (West Fort). These four hilltop artillery sites could effectively control the key access routes into Pretoria and they were within range of each other for mutual support.

Fort Hendrina, Makhado.

The Boer forces also dabbled in the concept of blockhouses when the Zuid Afrikaansche Republiek imported three prefabricated steel plate blockhouses from Austria in 1886. Movable and assembled on site, they provided a capability to defend troops or police from small arms fire. During 1887 the one shown here was used to protect 25 artillerymen and 100 mounted troops. Three of these structures were procured and the surviving example, named Fort Hendrina, is situated in Louis Trichardt, now Makhado. The fort is situated in the town's Erasmus Street, next to the library, and is a listed National Monument.[127]

During their life, these forts could be moved from site to site and were given different names and even modified in size, depending on the site requirement. Fort Hendrina, donated to the local municipality in 1969, was named after the wife of General PJ Joubert and was on the farm Klipdam (near Polokwane). It is not known what became of the other two forts.

Fort Hendrina measures 9.4 metres across by 3.75 metres high to the eaves, with loopholes for defence at the lower level and larger openings higher up for light and ventilation, and with a steel plate roof of umbrella form.[128] During the Second Anglo-Boer War the fort was captured by the British and renamed Fort Edward (after King Edward, who succeeded Queen Victoria in 1902). Breaker Morant and the Bushveld Carabineers operated from this fort in their effort to 'clean up' the Soutpansberg (salt pan mountain) area of Boer commandos.

The British Army and particularly the Corps of Royal Engineers had a great deal of knowledge and experience of defensive structures by the time the war had started. It was of course an army of manuals, drill and procedures, keeping everything recorded and tasks easy to accomplish for the average soldier. At the commencement of the war blockhouses were defined in the *Manual of Military Engineering*[129] of 1894 as follows: 'Blockhouses are defensible guard-houses or barracks, having framed or stockaded walls and roofs formed of timber, or iron rails, with earth on top. Being easily destroyed by artillery, they are chiefly suited to mountain warfare in wooded country, where it is not always easy to bring artillery to bear upon them; and where the artillery itself is of inferior power.'

It is an interesting definition and one which really did not stand the test of the first engagements with the Boers in the South African landscape, although by the time blockhouses were being constructed the majority of the Boer artillery was either destroyed or captured. As a static component of defensive warfare they become an easy non-movable target for long-range direct or indirect artillery fire, but only if you have artillery. With no artillery with which to engage, an enemy rather has to attack a well-defended structure with infantry and small arms, or avoid it completely.

DEVELOPMENT OF THE BLOCKHOUSE STRUCTURES

The British therefore came to the Second Anglo-Boer War with some experience of constructing fortifications across the Empire in recent but more widely dispersed conflicts, such as that in Afghanistan. In addition, a colonial history across the Americas was rich in the construction of frontiersman forts, stockades and blockhouses. Finally the ongoing conflicts in Cuba had demonstrated to the officer corps at least the potential value of blockhouses, notwithstanding the breadth and depth of knowledge possessed by the Royal

Engineers, several of whom were notable generals serving in the conflict. It is hardly surprising that when required the blockhouse was reborn in the conflict against the Boers as both masonry and lightly protected corrugated iron and mobile blockhouses.

Ad hoc fortifications and development

Fortunately the British force in South Africa was blessed with a plethora of Royal Engineers, and it was Kitchener (also a former Royal Engineer) who on assumption of the South African command, sought to upgrade the defences in order to better protect the lines of communication. General Roberts was already aware of the need for more robust defences and it was in September 1900 on returning from a trip to Komatipoort in the east that the first use of the term 'blockhouse' was recorded.

The Jones' Iron Band Gabion.[130]

Jones' Iron Band Gabion.

A 'gabion' (a French word derived from the Latin for cage) is defined as a wicker basket filled with earth and used to build defensive fieldworks, such as revetments or parapets.[131] This particular design was invented by Sergeant-Major J Jones, RE and was banded with galvanised sheet-iron strips 3.25 inches (9cm) wide and 77 inches (2m) long for extra strength.

In December 1860 the Royal Engineer Permanent Committee at Chatham and the Ordnance Select Committee gave directions that the new design be adopted generally in the service. The design was even showcased at the International Exhibition of 1862.

The old design of solely wickerwork took three men three hours to make, whereas Jones' gabion took two Sappers only four-and-a-half minutes. One hundred men could therefore construct 5,400 of the iron-band gabions in nine hours, while the same number of men, in the same time, would only make 100 wicker-only gabions. The other merits of Sergeant-Major Jones' invention were that they were more portable, lighter in weight, cheaper to make, simpler in their construction, more durable and fire resistant.

As an example, the total number of gabions used by the English at the siege of Sebastopol was 19,322, and had iron-banded gabions been used they could have been constructed by 100 Sappers in four days. In addition they could also be used as static points spaced apart on which to construct a barbed wire entanglement, to protect against infantry or cavalry advances. Winston Churchill even mentions the Jones' Iron Band Gabion in his book *My Early Life: 1874–1904*, and that his experience of practical fieldwork was 'most exciting' – maybe after a glass of champagne?

In August 1900 Major-General Elliot Wood (Engineer-in-Chief) had initiated some work to strengthen the sandbagged sangers with steel doors, similar in nature to those he had used in Suakin (Egypt) in 1888. At the same time Pretoria was being made more defendable with a ring of defensive points being built, such as Johnston's Redoubt which was noted by the Pretoria Commander Royal Engineers (CRE) 'as being well advanced' by December 1900.[132] By January 1901 the first iron and stone-filled wall blockhouses were being constructed by civilian contractors out of Lourenço Marques. These first experiments with fortified defences were rectangular, costly and difficult to build, and would be very similar to that currently still standing in Barberton, which is listed as a National Monument. Train wrecking was a significant problem along the Pretoria to Komatipoort line and this area became the development ground for hastily constructed fortifications. Major Spring Rice (RE), the pioneer for light prefabricated blockhouses, commanded the 23rd (Field) Company RE based in Middelburg.

EVOLUTION OF THE BLOCKHOUSE SYSTEM IN SOUTH AFRICA

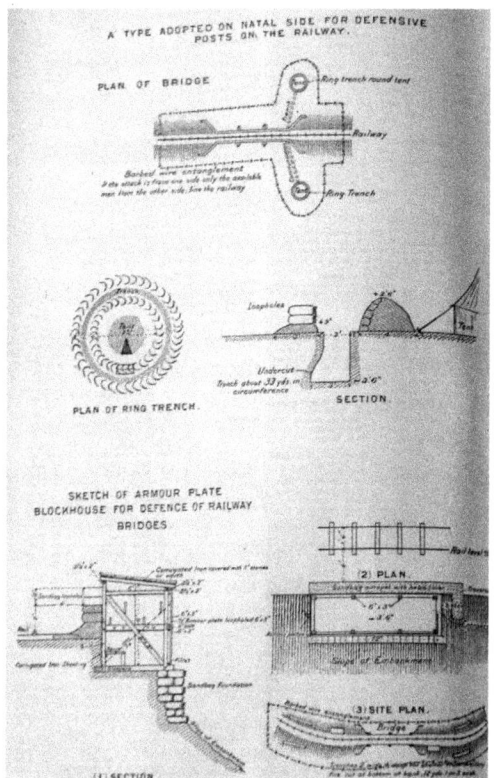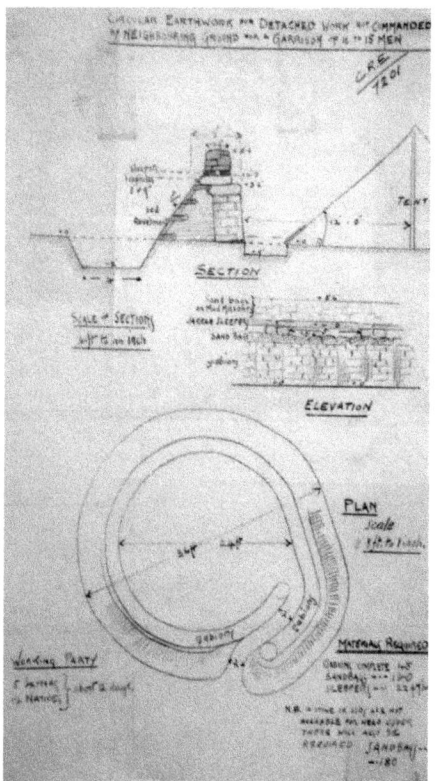

The types of early fortifications used to protect the rail lines prior to the introduction of more formal stone and corrugated-iron blockhouses. Left[133], early plate blockhouse and tented earthworks used to protect the key rail bridges. Right, details from CRE reports illustrating the use of wicker and steel caissons to build defensive walls from earth and shingle filling.

By the time the war was over it is estimated, from the author's extensive research into the records and from observations on the ground, that there were approximately 10,000 blockhouses constructed. This number would have included all other fortifications constructed such as 'works' and South African Constabulary (SAC) police posts, but excludes the numerous Town Guard forts, the small blockhouses in the passes and kloofs of the Karoo and surrounding mountains. This unaccounted number of forts for the 128 Town Guards alone could be as many as another 500 plus. For every category of design there is a hybrid or variation, created by the imagination of the builders or due to the availability of building material. The family tree shown here gives a holistic view of the key designs, and where found surviving examples.

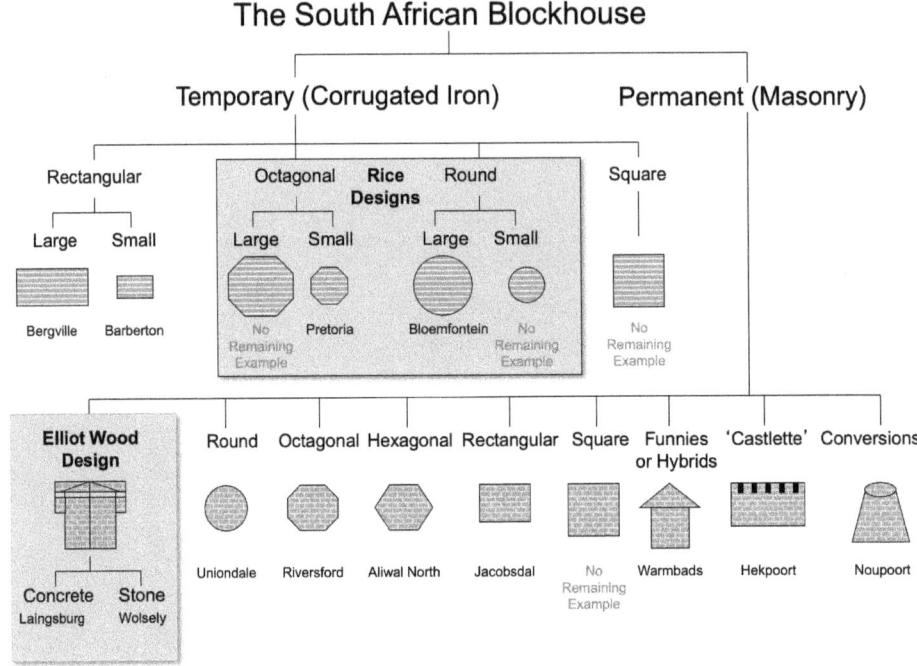

The blockhouse family tree.

MASONRY BLOCKHOUSES

Royal Engineer records note that 441 masonry blockhouses were constructed between December 1900 and the early part of 1901, though some sites were developed much later.

Elliot Wood Pattern Blockhouse

Many of these masonry blockhouses were built to the standard three-storey design by Major-General E Wood (Chief Engineer) using mortared stonework or unreinforced concrete; and were erected at important points such as railway bridges, railway stations, and towns and to guard strategic assets. It is estimated that using trained stonemasons and artisans they would have taken approximately 22,000 man/hours to construct over a period of 3–5 months. The records state that they cost between £800 and £1,000 to build and were built by a mixture of soldiers and contractors under Royal Engineer supervision.

There were some slight variations to the plans and in three instances the blockhouses were constructed out of concrete, at Laingsburg, Krom River and Merriman. Two other concrete examples located at Sout River and De Wet have since been demolished.[134] The three mentioned have survived surprisingly well in remote locations exposed to the elements; most can be visited easily and are situated close to the N1 Highway in the Karoo.

It is not clear from the records exactly how many of this type of blockhouse were built but currently 29 remain in various states of repair with the vast majority situated in the Western and Eastern Cape. These were constructed in the former Cape Colony to protect the Cape Town railway line taking supplies to the troops in the north. The most northerly example along this line is situated at Modder River. There is a second cluster on the East London to Springfontein line, two each at Stormberg Junction and Burgersdorp, and one at Aliwal North. Two of identical design are located on the hills overlooking Harrismith, while the other two known blockhouses were slight variations at Harrismith Reservoir, which is in good condition, and Kommando Nek 1, which is now a dilapidated ruin. Several other potential sites have been identified where this design most likely stood in the Cape region, indicating that this was a design that satisfied the early requirement for key point defence of bridges and railway stations. These were also key to protecting supply lines when the Boers were invading the British-controlled Cape Colony towards the end of the war.

These blockhouses took months to build and required considerable skill and resources to quarry local stone, manufacture it into useable stonework and then construct the blockhouse. Both local stonemasons and military artisans were used for this purpose, and often a considerable local work force. CRE Pretoria Lieutenant-Colonel RC Maxwell RE notes that work done by contractors for similar blockhouses involved 15 contractors and 312 black labourers used on average! It is hardly surprising as the author calculates that for an Elliot Wood type of blockhouse some 220 metric tonnes of stone was used; this was slightly reduced for a concrete blockhouse to 195 tonnes. The structures that are more distant from those in need of building material have survived the best, and were constructed to a high standard of workmanship.

The standard three-storey masonry blockhouse housed between seven and 40[135] men, commanded by a subaltern (an officer below the rank of captain) or a senior NCO. The ground floor was used for storage and water tanks, the first floor as a living area and the second floor for observation over the area of interest, such as the bridge approaches. The gap between the eaves of the pyramid-shaped timber and corrugated-iron roof and the top of the parapet wall on the second floor was closed by canvas 'drops' during foul weather.

Steel machicolated galleries were placed at two of the diagonally opposite corners to allow flanking fire along the walls in case of an attack. They could also be used as a mounting platform for a machine gun. Access to the blockhouse was by long ladder to a steel stable-type door situated on the first floor, to prevent ground-floor entry by an enemy force. The ladder could be drawn up inside in an event of an attack. Large double trap-doors were situated in the middle of each wooden floor and access to the next floor was by means of a wooden ladder; often other staircases were added to the sides for easier movement than steep ladders. The interior walls and woodwork were limewashed to allow for increased light, preserve the timber and for easier cleanliness.

Elliot Wood Pattern blockhouse at Burgersdorp.

There were 44 loopholes over the three floors, numbered to allow for spatial awareness and for the speedy dispersal of each soldier to his post in an event of an attack. Often ranges were painted onto the key loopholes to indicate weapon sight settings to the most obvious points, such as the end of the bridge.

The roof was fitted with galvanised gutters, which discharged rainwater through internal downpipes to circular corrugated iron water tanks on the ground floor. The tanks were topped up regularly in dry weather by train or water cart if supplies were low. Original plans show these water tanks located externally, but most remaining blockhouses show evidence of two internal 60 gallon tanks being used, allowing for better protection and ease of access. Food, water, ammunition and mail were also delivered by train or cart, often on a daily basis, depending on location.

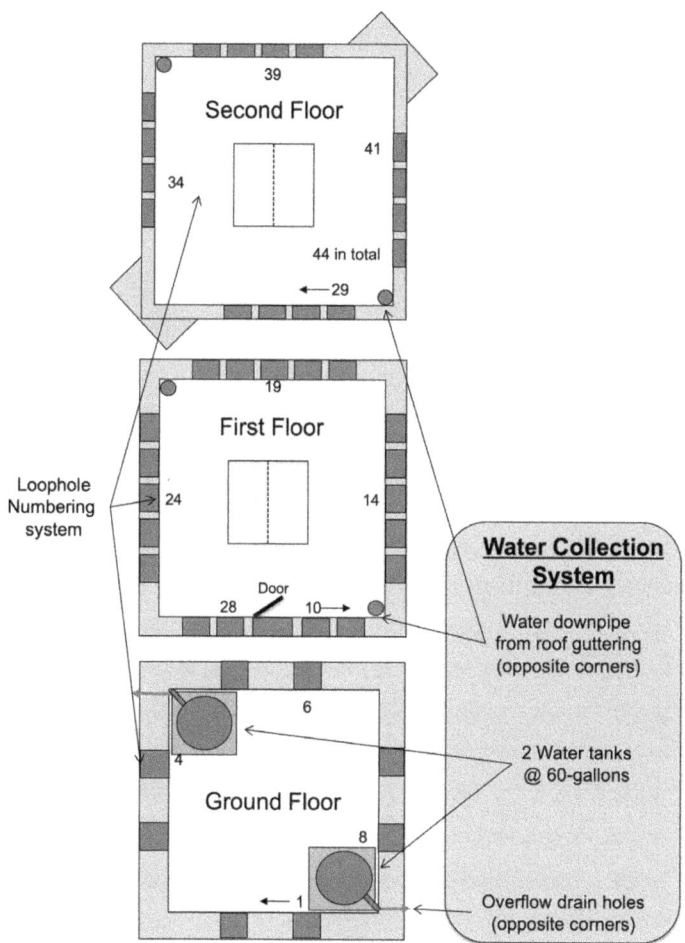

Floor plans of the Elliot Wood Pattern blockhouse.

No original design drawing exists in the Royal Engineer records, which also hold the private papers of General Elliot Wood. Only those diagrams in the post-war report, by Colonel Bethell are available and it is unclear if these were the original drawings or drawn retrospectively for the report. Once construction commenced however it was clear that improvement could always be made. A report made by Colonel Morris CRE of the Cape Colony suggests several improvements. The containment of the water tanks internally was generally accepted as 'safer and more convenient and can be more easily washed out'. This improvement to the design must have been made very early in construction as the majority of blockhouses show the drainage holes in the walls placed during the building phase and not afterwards.

The advice to add a canvas drop into the eaves on the top floor 'to keep out driving rain' also seems to have been adopted but is only evident largely in contemporary photographs. The only proof seen on site visits to all surviving blockhouses was in Jacobsdal, where they are evident. The suggestion was also made to add some sort of flap or cover to the loopholes to 'prevent draughts and to prevent unused loopholes being fired into by the enemy'. No evidence was seen at any of the blockhouses visited and it is presumed that the external low wire entanglements prevented the enemy getting close enough to be able to fire inside. Soldiers if cold no doubt found a way of blocking the holes at night to keep warm. The author spent a night in the Stormberg blockhouse which had modern day Perspex fitted over the loopholes, which certainly made for a dark and very cosy night's sleep.

Round Blockhouses

Constructing stone building to a round shape, while good for defensive effect, is difficult to achieve for the stonemason, and there are very few round blockhouses built from mortared stonework. Several of the Town Guard forts were built of dry stone work construction, which better lends itself to quick construction of readily available local and loose stone. Often built in or on the edges of small towns, they were a ready source of building material once the war was over, and few survive today.

One of the best examples at Danielskuil is constructed of the local ironstone and has a unique form and entrance way and was one of several surrounding this isolated town to afford it protection from the invading Boers on commando. In keeping with the importance of these structures, this fort and the small town fort at Uniondale are afforded National Monument status, though sadly they are both falling into disrepair. Other round structures have survived at Bakleikraal, Aties Fort and Krakedouw, all in the Western Cape, but are well off the beaten track and can be visited only after a hike.

A rare example of a round stone blockhouse between Irene and Johannesburg, now demolished. It shows a crenellated battlement observation floor, first floor entrance and six brattices provided for enfilade firing down the walls. It could be either one three storey built at Kaalfontein (Midrand) in April 1901 or the Zuurfontein blockhouse now commemorated by Blockhouse Street in Kempton Park.[136] Photo reproduced by permission of Durham University Library and Collections.

Octagonal Pattern

Covering all the approaches to a blockhouse is vital: once shut inside the occupants need to be secure in the knowledge that no-one can approach in dead-ground or blind spots, hence the development of the octagonal blockhouse. When a defensive structure has sharp ninety-degree corners, such as from a square design, either the entire corner is cut off to form an octagonal shape or it is chamfered in order to provide a portion of the wall for additional loopholes. There are two surviving examples, one at Norvals Pont, now converted to a private residence, and one at Riversford, located on private land 160 km to the north, close to a rail/river crossing. Several others built to an identical pattern and located in the same area along the Orange River, a pair at Norvals Pont to the north west of the bridge and the other somewhere in the vicinity, have been destroyed. It is not known who designed these blockhouses as so few seem to have been constructed.

Octagonal blockhouse at Norvals Pont showing small door platform and hoist. Courtesy of South African National Biodiversity Institute Archive. [137]

Of all the blockhouses built during this period these octagonal shapes seem to be the sturdiest and best constructed. The twin at Norvals Pont was deliberately demolished to construct a local farm and stables, and had it survived would have been an excellent tourist attraction. These were large structures (contemporary photographs show up to 20 men occupying them) and were built to either guard strategic assets or for use as headquarters, serving other blockhouses in the vicinity. Riversford, although renovated with a new roof, is still falling into ruin due to exposure to the elements and lack of maintenance. They are constructed to a larger scale than the Elliot Wood Pattern, and at about 6 metres square have a much grander appearance and no diagonal galleries. In this example the flanking protection along the walls is provided by a small balcony-type gallery on each flat wall. Occupying soldiers would have been able to provide a lookout from here and flanking fire down onto the walls if required.

Access was provided by a first-floor entrance and a system for winching up stores by means of an iron arm and hook onto a small protruding wooden platform. Internal sloping stairways are provided for easy access up and down and are most likely provided as part of the renovation to Riversford. The new external diagonal steel stairway is also an added feature to aid access; routine entrance would have been by a retractable ladder. The entrance is a steel stable-door with firing ports cut into it, which opens inwards and can be bolted and padlocked shut from the inside. There are 16 loopholes on this level and three larger steel-shuttered windows to provide extra light and ventilation.

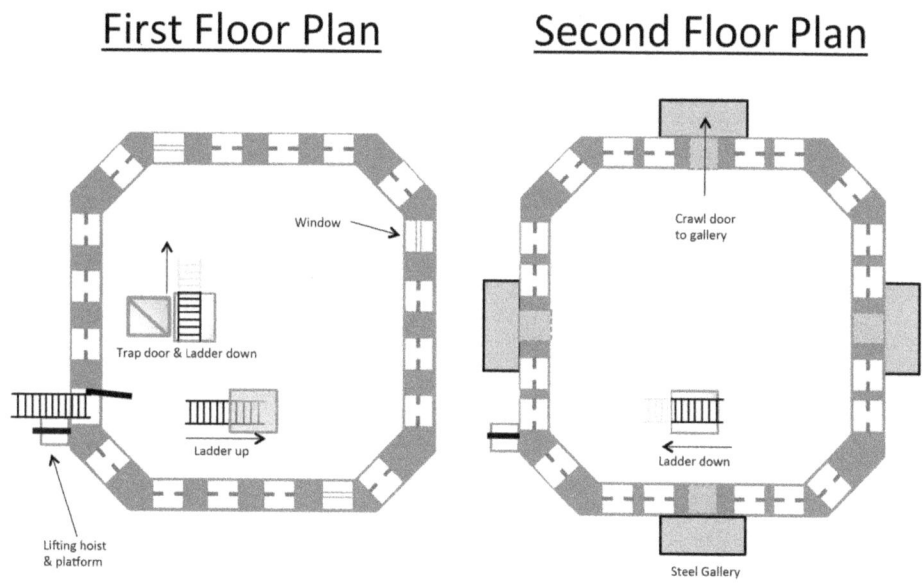

Floor plans of an octagonal blockhouse.[138]

The ground floor is provided with 16 loopholes for defence (three per side and one per chamfered edge). It is quite dark and would have been used mainly for stores. Access is gained by a small trap door and ladder; no internal water tanks are evident in this design of blockhouse. The floor joists are a mixture of wood and rail lines supporting a wooden floor which forms a fairly low basement ceiling. The first floor provides the key access route into the blockhouse and ladders to the other two floors. It is provided with another 16 loopholes, three larger steel-hinged windows and the main stable-door.

The second floor is accessed up a steel ladder and provides the highest point from which to observe the surrounding area. Each side houses a protruding steel-plated rectangular gallery supported on cantilevered rail supports. It is accessed via a crawl door and small steel doorway, which can also be bolted from the inside. The top floor is provided with 20 loopholes and four galleries. The roof is quite shallow pitched and has a small ventilation turret cover at the roof apex to provide some airflow. All the floors provide ample space to house men and material and could have housed up to 30 men. Each floor is provided with ample defensive options for all-round defence with about 20 firing positions per level.

Hexagonal Pattern

The hexagonal shape also proved popular with some blockhouse designers and two examples remain in various states of repair in Aliwal North. They were well constructed out of mortared stonework and have entrances at ground level, giving an indication that these were garrison blockhouses located in relatively well-protected areas, where the threat of attack was considered low.

Aliwal North is unusual in that there is also an Elliot Wood pattern blockhouse situated to protect the town. This indicates that there were most likely two phases of defensive construction, and that these hexagonal blockhouses were built at a later date once this area had been fully secured. The hexagonal design also requires much less steelwork but has no capability to provide flanking fire along the wall. The inside is unusual in that the loopholes are situated well over two metres above ground level and some type of banquette (step) would have been required along the wall to allow soldiers access to fire from them. These were most likely wooden and removable, as no sign exists in the well-preserved example in the Dewetville Street blockhouse today. From the outside the loopholes appear above the standard height doorway, giving a more imposing impression to the blockhouse.

This site is preserved as a National Monument and located in a well-kept residential suburb, unlike its sister blockhouse, which only remains as a stonework shell, located to protect the railway station and approaches. These blockhouses have two storeys, the upper one having a small but open space between the parapet and the shallow umbrella roof, providing good all-round observation and firing positions. No floors exist in the surviving examples, but it is assumed that a ladder and trap door arrangement would have been used.

Aliwal North 2, on Valk Street, Aliwal North in 2012.

Rectangular Pattern

There are also small blockhouses built to a rectangular pattern, and two main concentrations existed, one with blockhouses at Jacobsdal and Warrenton and the other around Waldrift (Klip River), north of Vereeniging. The contemporary photographs show the double-storey variants to have been situated around Vereeniging, while the single-storey versions were in the Warrenton area. Both examples often had a small observation tower built at one end to provide a better view over the surrounding area. The entrance was at ground level and they were quite roughly built from local stone, sometimes stone carved, but otherwise from loose surface rocks, mortared into walls.

Waldrift (Klip River) blockhouse, now demolished.[139]
Courtesy of SANDF Photo Archive.

Square Patterns

Square blockhouses were also popular and many were built in the same style as those rectangular patterns with the small towers. At least two blockhouses of this type were constructed at Vereeniging, close to the Vaal River rail bridge, now demolished. There is one surviving example at Witkop just south of Johannesburg on the A57 highway which has some interesting additions in the form of angle bastions and is strikingly similar to the Klip River blockhouse shown here; indeed these two sites are only 17 km apart. Their geographical classification as the 'Vereeniging Pattern'[140] although not adopted by the author is valid and confirms that regional units or more likely Royal Engineer companies played a significant role in determining what shapes and designs were adopted in their areas.

Square blockhouse[141] in the Brandwater Basin of an enhanced 'Brindisi Pattern', having an additional stone lower level added, most likely for enhanced fields of fire.

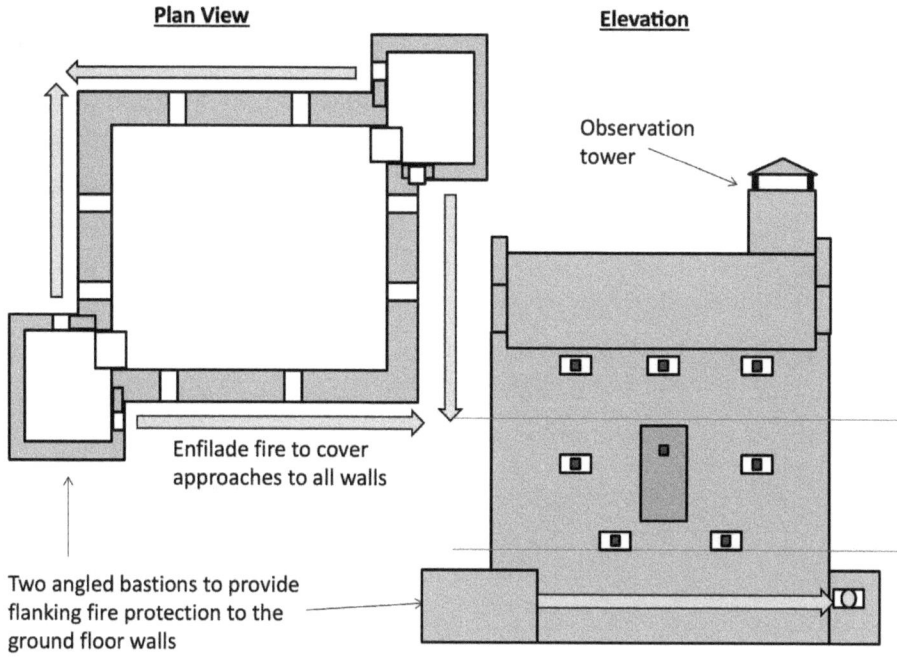

The Witkop Blockhouse with its added angle bastions.

Castlettes

The saying 'An Englishman's home is his castle' applies very aptly to blockhouses built by the British. This was reflected in the seminal work of Richard Tomlinson, *Britain's Last Castles: Masonry Blockhouses of the South African War, 1899–1902*, which baselined the state of South Africa's blockhouses in 1995. There are several examples where the blockhouse designer must have drawn on a medieval vision of the crenellated battlement of the castle ramparts. Crenels (spaces) and merlons (solid stone projections) provide the typical castle outline everyone remembers from castles such as Windsor Castle or that belonging to the Sheriff of Nottingham. The cover illustration of Fort Campbell is the best found in surviving photographic records, having well-defined and evenly spaced crenels and merlons on the battlements and a small turret for greater observation or in this case the symbolic flagpole complete with Union flag. Several of the other designs also take some type of crenellation into their top floor, topped off by a corrugated-iron roof. The type classified here has a flat roof open to the elements.

In terms of their geographical location the majority survive in the area around Johannesburg, at Krugersdorp blockhouse and in the Magaliesberg at Hekpoort. Here the passes over this range to the west of Pretoria attracted several different design types in Kommando Nek, Hartbeespoort. Little remains of the four identified blockhouses that guarded the pass and the surrounding hills, but contemporary photographs show square, 'T', 'L' and 'X'-shaped designs, all with crenellated roofs. They were substantial structures in their remote areas, where materials were brought up by packhorse or temporary cable lines. There are several well-constructed mule tracks still evident in the Magaliesberg, with dry stonewalls and revetments still in use by hikers to this day. Cement seems to have been in short supply for these blockhouses, as local soil with little added cement has quickly led to their collapse from the elements and vandalism.

Fort Campbell blockhouse, occupying a commanding position in the Brandwater Basin.[142]
Reproduced by permission of Durham University Library and Collections.

This type of blockhouse was also provided with loopholes and shuttered windows and often had chamfered corners to provide covering fire into any potential dead-ground. The single roof was often constructed using supported arch sections of corrugated iron onto which concrete was poured to form the top observation level. Doors were of thick steel plate seen most often in most other of the blockhouse designs, .

Hybrids and One-offs

Hybrid blockhouses demonstrate some kind of adaptation or joining of two designs together, whereas one-offs are just that, and typically unique, not fitting into any particular classification.

The Prieska blockhouse is a fine example, hexagonal in shape but having a unique battered wall, sloping so as not to allow potential attackers close to the wall. The crawl entrance and open roof make this a stand-alone example, and it stands a resolute testimony to those who constructed it out of local Tiger's Eye stone on the town's prominent hill.

Johnston's Redoubt in Pretoria was constructed in a natural pass over Meintjieskop, a prominent ridge to the north of the main old town. It was one of the earliest blockhouses constructed in 1900 on a farm called 'Lisdogan' as the British occupied Pretoria and before Elliot Wood came with his prominent deign. The square single-storey redoubt did have five-sided fan-shaped galleries at opposite corners, but was not using the long cantilevered rail lines as seen in Wood's design. This version is large at 6.2 metres, and is provided with a door at ground level, loopholes and shuttered windows on three of the remaining walls. Access to this blockhouse is sadly not available to the public,[143] as it is now protected inside the Presidential grounds. It was most likely named after Andrew Johnston, the merchant who owned the farm Lisdogan. He originally came from County Sligo in Ireland, and later twice became the Mayor of Pretoria.[144]

Johnston's Redoubt in Pretoria. Courtesy of SANDF Archives.

Danielskuil in the Northern Cape is another blockhouse (or Town Guard fort) in a category of its own, and a protected National Monument overlooking this remote Northern Cape town, once famous for mining lime. It is a small circular single-storey, dry-stone construction built of orange-coloured slate-like slabs of local ironstone, with a protected crawl entrance and no discernable roof. It is the only survivor of five constructed to protect the town from Boer incursions into the Cape Colony.

Some oddities[145] (1) Hexagonal blockhouse with battered walls and crawl entrance and pitched roof (Courtesy of SANDF Photo Archive);
(2) Vredefort Road Koppie blockhouse (Reproduced by permission of King's College London Archive), with corrugated-iron galleries and a crenellated open roof;
(3) Danielskuil blockhouse / Town Guard fort built from local ironstone slate;
(4) Rectangular corrugated iron blockhouse on a pile of sleepers, presumably to give better observation and field of fire
(Reproduced by permission of the National Army Museum, London).

Conversions

The British of the day were good innovators and adapted many existing structures, fortified them and used them effectively as blockhouses. The first of these conversions took place during Roberts' command when the key stations along the supply lines were fortified with sandbags, trenches and wire. Many of the ganger's – or foreman's – huts along the rail lines were also fortified and used as bases or blockhouses. These huts appeared on several of the maps produced at the end of the war by Major Rice showing the defences of the Pretoria to Komatipoort line. As the war progressed, some of the farmhouses and buildings situated conveniently for the British were also fortified. The Boer farms were generally well built and once loopholes were cut often made sound defensive structures.

One of the most interesting conversions is the blockhouse at Noupoort, which could not have been anything other than a windmill, and is located close to the railway line on Windmill Hill. It resembles many of the existing Cape Town windmills in shape and form, but all of the features such as windows and doorways seem to have been blocked off and plastered over. No contemporary photographs have been found in archive searches despite this being a major British railhead during the war. It is a tall structure at over 7 metres high and today the only visible doorway is at the gable end of the roof. Climbing up a ladder that high would have been precarious, to say the least. The loopholes for firing and lookout are small and observation would have been difficult from anything other than an open roof. Other uses –because of the blocked-off entrances – may have been as a telegraph station or storage facility. It is impossible to access so the internal configuration is unknown.

Noupoort blockhouse, a former windmill.

CORRUGATED-IRON BLOCKHOUSES

The need to protect the rail lines from Boer train attacks late in 1900 had already prompted the Commander Royal Engineers (CRE) Major-General Elliot Wood to refine his ideas from the structures he had been responsible for building in Suakin (Egypt) 15 years previously. In his memoirs he states that the sandbagged, loopholed and steel-plated door design was his, although the shape bears little resemblance to what was built in Egypt. The Suakin redoubts were however built for a different purpose. They required height for the signal tower and most likely drew from the skewed top-storey approach adopted in many American frontier forts to gain the best protection. The need to improve the rail protection from purely sandbagged trenches into fortified structures resulted in the first blockhouses being constructed by civilian contractors from Lourenço Marques (Maputo) on the eastern edges of the occupied territory around Nelspruit, Kaapmuiden and Komatipoort, where there were long and exposed rail-river crossings. The need for something other than trenches and sandbags was well appreciated throughout the Engineer Command and Major Spring Rice RE of the 23rd (Field) Company RE in Middelburg notes in his memoirs 'that defence by means of open works was only asking for trouble'.[146] He was involved in contracting for these early blockhouses as they were in his Company's area of operations after the occupation of Pretoria. The Royal Engineer Boer War diary notes that they ordered masonry blockhouses from Army headquarters to be employed as defence posts to add to security and reduce as far as possible the use of troops for fixed defence. The problem was that train attacks were on the increase and the line from Pretoria to Komatipoort was extremely long; an alternative solution was required. Previously only open works had been used to guard stations and bridges, using corrugated-iron sheet and earth filling; the time was right for something more substantial.[147]

Early Designs

The early blockhouse designs were rectangular and built with interior and exterior wooden stays dug into the ground as two parallel rows to support the corrugated-iron walls 2 feet apart (610 millimetres). The space was filled with stones. Holes were then cut into the walls, most likely in situ to provide openings for the loopholes. These were steel plates fitted into wooden frames, which were then inserted into the part-filled walls, and the remaining filling poured in. Walls normally formed a rectangle, but in some cases as shown in the Komatipoort blockhouse below, they were curved to accommodate the tactically more advantageous rounded corners, with an additional loophole. This design alleviated the threat caused by the 'blind-spot' at the corner approaches to the blockhouse. The roof was single-skin corrugated iron and a door was provided in one of the long sides. The whole blockhouse was then surrounded with protective barbed wire and a water tank added inside this wire entanglement.[148]

Millicent Garrett Fawcett.[149]

Born Millicent Garrett in Aldeburgh, 1847, as a suffragist she took a relatively moderate line for the time, but was a tireless campaigner, concentrating much of her energy on the struggle to improve women's opportunities for higher education, after the death of her husband.

In July 1901 she was appointed to lead the British government's Fawcett Commission to South Africa and investigate conditions in the concentration camps that had been created there. Her report corroborated what the campaigner Emily Hobhouse had found on her investigations.

An early 'modern blockhouse' as Millicent Garrett Fawcett noted in her album,[150] most likely taken in the Komatipoort area on her tour of the country. Courtesy of the London School of Economics.

Only about six of these early blockhouses were initially built, most likely constructed in pairs at the respective river crossings along the Komatipoort rail line. They were a good strong design, but time-consuming to construct; moreover, at the time steel plate for the loopholes was not quickly obtainable or in sufficient quantities to allow for any sort of mass production. The armoured steel-plated traction engines were already being stripped of plate on arrival in-theatre in order to armour the trains.

The mechanical rolling machine that made the Rice Pattern blockhouse possible. This rolling machine (operated here by the author's son) is in the the War Museum, Bloemfontein.

The Corps of Royal Engineers was already heavily committed in the theatre of operations, on the railways, repairing bridges, constructing depots and remount centres for the horses, and drilling numerous boreholes for the provision of water. At the forefront of this was the team of officers commanding 23rd (Field) Company RE, Officer Commanding Major Spring Rice RE and Captain George Henry Fowke RE, his second-in-command. Already experienced from their entrapment at the siege of Ladysmith, these engineer officers found themselves stationed in Middelburg, east of Pretoia, where they stayed for the remainder of the war. The British, now commanding the towns and securing the railway lines, found themselves easy prey to Boer sniper attacks. Boer 'hunting' opportunities were created by cutting telegraph lines and then sniping at the repair party, or by engaging the raised signalling towers at long distances from good cover. Protection of these sites fell to the Royal Engineers and it seems George Fowke first devised the concept of forming sheets of corrugated iron into simple walls filled with stones to provide protection to these vulnerable signal sites in the last quarter of 1900.[151] The concept of caissons was already well used at redoubts, forts and for gun emplacements, where strong wicker cylinders had already been used filled with earth and stones and supplemented with sandbags. They were however very large and designed to protect against artillery fire and would be overkill for small-arms protection. The concept of thinner stone-filled flat walls for protection from rifle fire was a new idea.

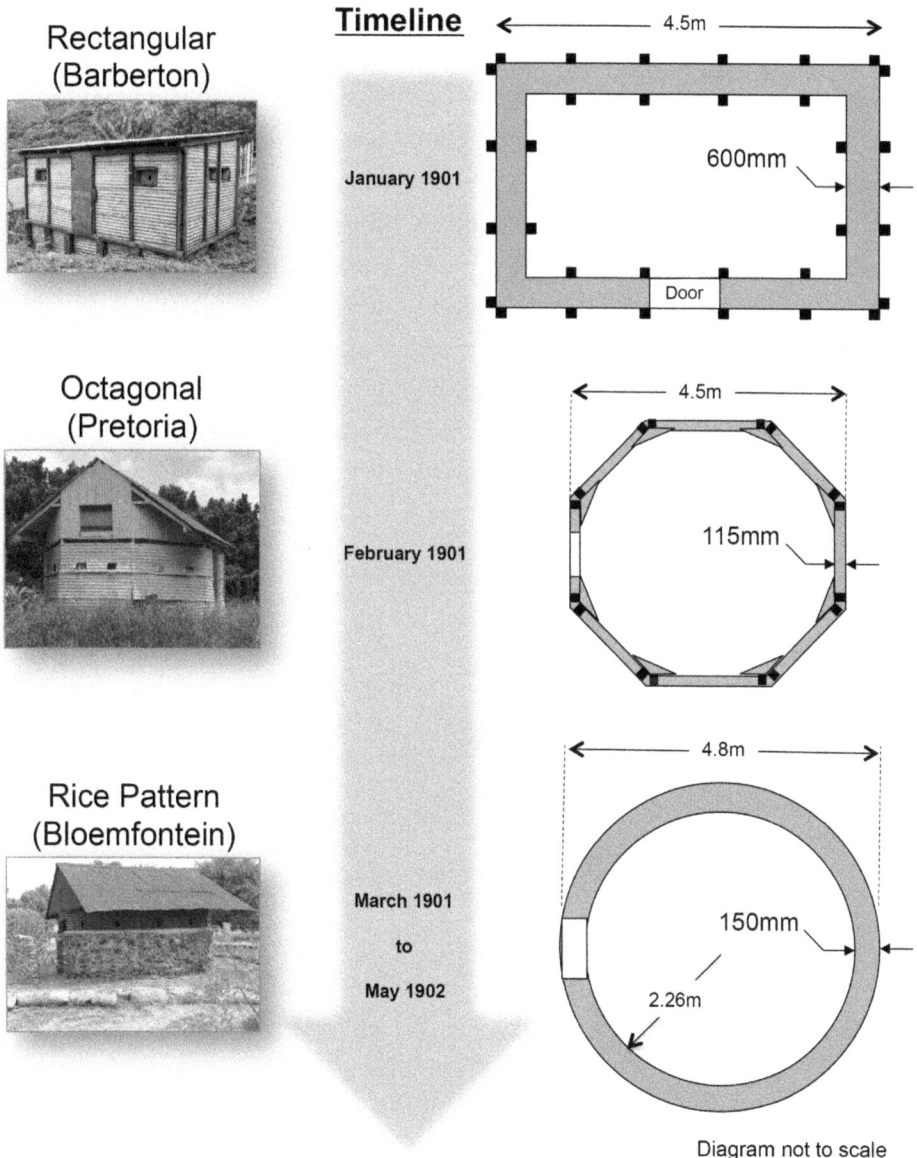

Evolution of the corrugated-iron blockhouse.

Octagonal Blockhouses

It was in January of 1901 that the first corrugated-iron shingle-filled octagonal blockhouses were erected. After extensive field firing tests the walls were much thinner at only 4½ inches (115 millimetres); seven sides of the eight-sided octagon could be prefabricated, and apertures cut for the double funnel-shaped sheet-iron loopholes, obviating the need for steel plate. Small holes were drilled into the loophole sill, so that additional infill could be added under the loophole in cases where the filling had settled. There was always the danger that the wall did not protect sufficiently if this settling occurred, and several men were shot in this manner.[152] In addition, the corners were mainly wood and needed additional protection across the internal corners, and the walls were prone to bulging outwards and becoming weak.

For construction all that was required was for the ground to be levelled to a firm foundation, usually on bare stony ground, and then the eight sides, which were bevelled at each corner, slotted and bolted into place. The eighth side housed the doorway, which was impossible to make from corrugated iron and shingle and had to be made of wood, and the base was topped off by a high square gable-ended roof. The full-size wooden door was a significant vulnerability to rifle fire and this was either replaced with a low crawl entrance or protected with a large revetment blast wall.

An early octagonal pattern blockhouse showing the additional stone revetment and the thin CGI and shingle blast wall to protect the doorway.[153]
Reproduced by permission of the National Army Museum, London.

Concern was initially raised on the 3/8 inch (10 millimetres) thickness of steel plate supplied from Johannesburg being used that it might not give protection against the Boer Mausers. Tested at a range of 70 yards, this proved to be the case, and additional ¼ inch (6 millimetres) backing plates and ½ inch (12 millimetres) plates were accordingly obtained as reinforcement.[154]

On 31 January 1901, the first CGI blockhouse was erected in defence of the railway bridge at Klein Olifants, 3 miles (5 kilometres) east of Middelburg. It was octagonal in shape with 16 loopholes and was the prototype of many others of this kind.

The octagonal design was popular and a production line was quickly established by Rice to mass-produce the template-framed walls and components in kit form. This pattern appears in most areas in contemporary photographs and originated from the Middelburg-based Engineer Company. During the first month of production in February 1901, 50 blockhouses were delivered from Middelburg, including some of the rectangular design for bridge protection.[155] The octagonal blockhouse design of Rice's team was acknowledged by Kitchener (also a former Royal Engineer!) and quickly exploited to provide protection along the rail lines in most of the other areas.

Round Blockhouses

It was in March 1901 that Major Rice took delivery of a corrugated iron rolling machine for manufacturing water tanks, another important engineering task. This gave him the idea, perhaps fuelled by the sight of African

Blockhouse Development Timeline

1900

5 June 1900
Roberts occupies Pretoria – issues orders to fortify lines of communication with sangars and trenches at key points (stations, bridges and culverts).

July 1900
Key points fortified with sangars and trenches completed – train attacks continue.

August 1900
Rail protection Delagoa Bay line: Elliot Wood initiated ad hoc sandbag blockhouses with steel-plated doors, similar to those he had deployed in Suakin, Egypt.

September 1900
Roberts returns from Komatipoort – first use of the term 'blockhouse' in theatre.

December 1900
The CRE, Pretoria District, records that Johnston's Redoubt, East Fort, and Howitzer Redoubt were 'well advanced' by 1 December 1900. Train attacks reach peak.

1901

January 1901
Further protection of Delagoa Bay rail line: use of stony sand to infill the space between corrugated iron walls of rectangular blockhouses. Built by Lourenço Marques contractors.

January 1901
First blockhouse developed by Major Spring Rice: Octagonal Version 1, corrugated walls filled with shingle, built at Klein Olifants River. Initial ideas used widely by other Engineer Companies, usinf their own manufacturing methods.

traditional rondavels, to make a round blockhouse. One of these machines used during the war is sited close to the Bloemfontein Rice Pattern blockhouse in the War Museum, Bloemfontein. By manufacturing several sheets of a certain curved radius several could then be bolted together, without a supporting wooden framework, to produce a fully circular blockhouse. Two light separated cylinder walls could then be filled with similar graded stone and shingle available on the veld to provide adequate protection against small -arms fire.

The design had moved to its final stage and last iteration: now there was less wood required for the walls, and only small wooden spacers were needed to ensure the correct separation. The wall thickness could now be increased to 6 inches (150 millimetres), and there were no vulnerable wooden-framed corners. The doorway was often a low crawl way, but in some areas was a normal full-sized door, protected by a stone or sandbagged blast wall. Major Rice had admitted to being influenced by the low entrances of local African huts, and this entrance was designed to prevent the enemy rushing through an obvious door-way. Internally there was a small shield of corrugated iron as an anti-blast wall in case of grenade or small-arms fire.

The roof again varied according to construction teams and areas, some being a tall gable-ended version and others the lower-profile octagonal umbrella roof, the former being cooler in the summer, the latter warmer on the winter nights. The tall square-gabled roofs also required strong fixing against the wind and often this required tying down with

13 January 1901
First deployment of blockhouses: Komatipoort area – six rectangular corrugated blockhouses built for rail protection, by contractors from Lourenço Marques.

February 1901
Rice quickly developed the octagonal version into a quicker and easier to construct Version 2.

March 1901
Major Rice takes delivery of a corrugated-iron rolling machine, designed for manufacturing small water tanks.

11 March 1901
Rice developed the circular Rice Pattern corrugated blockhouse due to the demands for rapid mass production at low cost. The first was erected at Gun Hill, Middelburg. First factory established at Middelburg

March 1901
Elliot Wood Pattern masonry blockhouses start to be built at key rail points and into 1902, some 441 being constructed.
Rice type mass-produced and 'blockhousing' in full swing along the rail lines in the Transvaal and ORC.

April 1901
The first 50 blockhouses produced and built either side of the rail line at Middelburg.

June 1901
Kitchener deploys the first cross-country route Groot Olifants River station to Val station, linking the Komatipoort and Natal rail lines.

July 1901
Reinforcing of the rail line in the Cape Colony to counter De Wet's intended invasion of the Colony.

wire stays into fixed posts or very large rocks. Replacement of the square-gabled roofs by a circular one gave rise to the name of 'Pepperpot Blockhouse'; in Northern Natal[156] this design was an octagonal ribbed roof referred to as the 'Ladysmith Pattern' or 'Burnaby design'. In some instances the roofs were initially deployed with a canvas tented arrangement with a central pole. This was usually a temporary measure, implemented to reduce the logistics burden and speed of construction. Whatever the design modifications were, this new design for light prefabricated blockhouses was much more economical and effective than previous ones. The 'Rice Pattern Blockhouse' was born.

A Rice Pattern blockhouse near Lindley, Orange Free State, built for speedy occupation with a canvas roof only.[157]
Courtesy of The National Army Museum, London.

July 1901
Blockhouse drive phase: blockhouse lines extended across the open veld to create 'nets' or 'pens' into which the roving commandos could be driven.

September 1901
Johannesburg protected with a ring of blockhouses. Hekpoort blockhouse built.

November 1901
By the end of November, the area around Vereeniging was similarly protected.

December 1901
Longest blockhouse line started from Victoria West to Lambert's Bay, and was completed by the end of the war.

1902

January 1902
First steel-plated 'travelling' blockhouse built on a four-wheeled ox-wagon, was used for convoy work, blockhouse line protection duty and to reinforce places where the Boers may cross existing lines.

March 1902
Experimental design of placing a small round Rice Pattern blockhouse on ox-wagon was developed, which could be used for mobile protection but only on hard roads.

May 1902
Cape Colony blockhouse line still under construction.

By the signing of the Peace Treaty of Vereeniging at Pretoria on 31 May 1902, according to official records over 8,000 blockhouses had been constructed, covering about 3,700 miles (6,000 kilometres), at a cost well under £1 million.

The initial drawings however were either too vague or the notion of the two different-sized cylinders being self-supporting was not grasped by all Royal Engineer units, and many were built to style but 'off-plan'. This is most likely the reason for several other variations! Small models of the design were constructed and sent from Middelburg to all units in order to get uniformity of build, which did eventually occur. Mass production then began and each blockhouse was manufactured as a complete kit, capable of ox-wagon transportation to any remote area.

Contemporary photographs of the 'Ladysmith Pattern' design typified by the unique arched loopholes made from corrugated-iron sheets as reported by the CRE Natal in 1902.[158] Reproduced by permission of the National Archives of the UK.

The blockhouse however was only the 'keep' in medieval terms (that is, the safest and strongest part of the castle, usually at the centre or on top of the motte in a Motte & Bailey design): it also required protection with an additional battered wall, a spider-web of barbed wire, a perimeter fence, and a sentry trench 4 feet 6 inches (1.4 metre) deep. The water tank was usually dug in to protect it and was often sited in the sentry trench. This produced a mini garrison, which varied from seven to 15 men, usually commanded by a JNCO, and supported with a night-guard African contingent of three or four men. In addition, there were alarms to make, signal rockets to deploy and numerous other construction tasks to complete, which are all reviewed in Chapter 5.

Once all this work was completed the task of interconnecting one blockhouse to another began. Initial distances were cited as one mile (1.6 kilometres) apart; later in the war they were reduced to half a mile (0.8 kilometre). The trench-digging work was usually carried out by the black workforce and the wiring party consisted of soldiers. This took a great deal of time and there was a process of continuous improvement underway, such that all the blockhouse lines became stronger as the war progressed. The early occupation was a very busy period, with constant threat from Boer raiding parties, although boredom very quickly set in in the later stages.

Such was the utility of this new lightweight fortification that they were mounted on ox-wagons for mobile blockhouses, examined in Chapter 6. In addition, they were placed on the headgears of mines in the Witwatersrand goldfields, on the roofs of prisons, and even used as mobile quarters for officers on mobile column duty. The density (one every mile and a half) of the initial blockhouse lines was also regularly increased by inserting new blockhouses to reduce the distance to often less than half a mile, such was their flexibility.

Despite the circular design requiring much less material, especially wood, there was now a huge demand for corrugated-iron, which was supplied from the home base, much of it being produced in Wolverhampton, which arrived at the coastal ports in huge shipments. Now mechanically if not hand rolled, the sheets were ready for on-site assembly after their factory pre-fabrication. The transition from small rectangular fortification into octagonal and then round blockhouse had also delivered improvements in overall living area inside the blockhouse.

Examination of the plans Bethell produced after the war from a mathematical perspective reveals that there were actually two distinct methods used to construct the Rice Pattern blockhouse. The author calculated the quantity of shingle required for a full-height (6 feet / 1.85 metre) corrugated -iron blockhouse as seen in many contemporary photographs as 5.24 cubic metres. This leads to a discrepancy in the quantity of shingle the plans suggest, being higher by a factor of two. If the plans are read carefully there is actually a half-height revetment caisson or 'parapet' constructed first. A levelling gauge is then provided such that when the parapet was finished the shield was to be lifted bodily to its place and held level with the gauge until shield was filled with shingle. On this basis there was recommendation for 3 cubic yards of shingle to be provided, which converted to 2.29 cubic metres of infill.

If the shield placed on the revetment is half the full height this means that there were two different methods at least to construct these prefabricated blockhouses. Some were full-height double cylinders from ground to ceiling, and others were shields placed on top of a built revetment wall.

Pepper-pot variant of Rice's design (left) showing the half-shield fitted onto a dry stone revetment, also known as the Burnaby Type; and the standard Rice Pattern (right) with a full-sized corrugated-iron wall.[159]
Courtesy of SANDF Photo Archive.

Cost Effectiveness

The corrugated iron blockhouse design had mastered the art of low-cost prefabrication, using mass production techniques pioneered by Henry Ford in car production. Each man in the factory had only one job and turnover of blockhouse kits was now like clockwork in each of the factories, to feed the increased requirement now demanded by Kitchener. The cost of the Elliot Wood Pattern is noted as £800–£1000 and the comparison shown below illustrates how much more cost effective these smaller iron blockhouses were to produce, transport and assemble. Although developed for different purposes, it is evident that smaller mass-produced blockhouses were just as effective in defence but now could be used much more extensively.

The Boer artillery by this stage of the war had all been captured or destroyed, allowing for a much reduced need for protection against artillery. The Boers did have some guns after their own were lost, and guns captured from the British were used – though not against blockhouse lines – in various encounters. The last one captured was in the Transvaal in March 1902, when they captured four British guns.[160]

The Rice Pattern blockhouse is more remembered and recorded historically than the Elliot Wood Pattern, but due to its light and temporary nature, only about six original versions remain today. Many of the old lines can still be traced from the interconnecting trenches and from their cleared foundations and shingle spoil heaps, but over 6,000 Rice Pattern blockhouses were disposed of immediately after the war. They thus had a less visible legacy than the Elliot Wood Pattern, of which about 29 remain in various states of decay.

The corrugated iron blockhouse had moved in three months from an ad hoc difficult to construct rectangular fortified house into a well-designed and easy to construct purpose-built blockhouse, ready for mass production and prefabrication in kit form. The summary table overleaf shows the relative improvements made during this rapid prototyping design phase, and how the pendulum swung from initial designs having a lot of disadvantages to the final design with many advantages over its predecessors.

Rank	Mounted	Dismounted
Majors	1	
Captains	1	
Lieutenants	3	
Surgeons	1	
Sub Total	6	
Sergeants	1	6
Corporals	1	6
2nd Corporals	1	6
Shoeing Smiths	1	
Sappers		134
Drivers	26	
Trumpeters	1	
Buglers		1
Batmen		12
Sub Total	31	165
Grand Totals	47	165

At war strength a field company was commanded by a Major, with 5 other officers. It was established for 21 NCOs, 175 men, 15 riding horses and 31 draught horses, and 6 store wagons.
Source : RE Chatham Museum Archives website.

Distribution of Horses :
- Officers (Private) - 2
- Officers (Public) - 9
- NCOs - 3
- Trumpeters - 1
- Store Wagon - 24
- Pack Equipment - 3
- Spare - 4
- Wagon, store - 6
- Total 52

Small Arms
- Rifles - 152
- Sword bayonets - 1
- Cavalry Swords - 6
- Pistols - 1

Cartridges per man
- Rifles - 50
- Total in Company - 7,500
- Pistols - 24

A Royal Corps of Engineers Field Company Order of Battle (ORBAT), c1899.

Advantages	Disadvantages
Rectangular (Barberton)	

• Good thickness of defensive wall
• When compared to stone blockhouses
  o Cheaper to build
  o Quicker to build
	• Required a lot of materials.
	• Complex to erect
	• Time consuming to erect
	• Lack of steel plate for loopholes
	• Blind-spots at corners
	• Vulnerable to enfilade fire
	• Lacked ventilation, and hot to operate from
	• Relatively small

Octagonal (Pretoria)	

• Could be prefabricated and industrialised
• Less material to build
• Provided all-round defence
• Easier to ventilate with gable roof, cooler to operate from
• Thinner wall meant less shingle
  o Cheaper to build
  o Quicker to build
• Much improved floor plan
	• Corners prone to weakness/bulging
	• Corner not bullet-proof
	• Complex to erect
	• Under-protected wall, rather thin
	• Full-height door not well protected nor easy to defend – required additional 'blast' wall

Rice Pattern (Bloemfontein)	
• Rapid & simple prefabrication • Much less material • Less complex; quicker to erect; • Self-supporting once erected; stronger with no bulging. • Very cheap to build • Could be made small or large • Adapted for lower profile roof – better to camouflage • Could be moved quickly to other locations, easy to transport • Uniformity aided the standardisation of tactical procedures • Developed for a movable (Ox-wagon) version • Larger floor plan	• Required a forming machine to round the wall sides • Lack of vision once inside (especially at night) • Initially vulnerable to fire from outside • Largely relied on fixed fire down the wire (especially at night) • High, pitched gable roof on cylinder, prone to wind blowing off

Relative advantages and disadvantages of each type of corrugated-iron pattern blockhouse (with existing example sites).

EVOLUTION INTO MASS PRODUCTION

The development of a small and effective mass-produced blockhouse was a solution to Kitchener's problem of how to catch the roving commandos. Now employing a more extensive 'area system' utilising the fortified railway lines as the backbone, he could add cross-country lines and move these at will to push out and secure more and more of the wide-open veld. At least that was the concept; the next step was to increase production. Several blockhouse lines were actually moved from one area to another as the demands of the war dictated, such was their portability.

Elliot Wood ordered each Engineer Field Company to establish a factory to supply its own area; these often had to send blockhouse kits to other areas requiring additional supply. Factories were also established at the ports of Cape Town and Durban, as all the the raw material was being shipped from Britain. The mass production process was now fully industrialised, initial design flaws corrected, plans drawn and models sent out to ensure conformity to design.

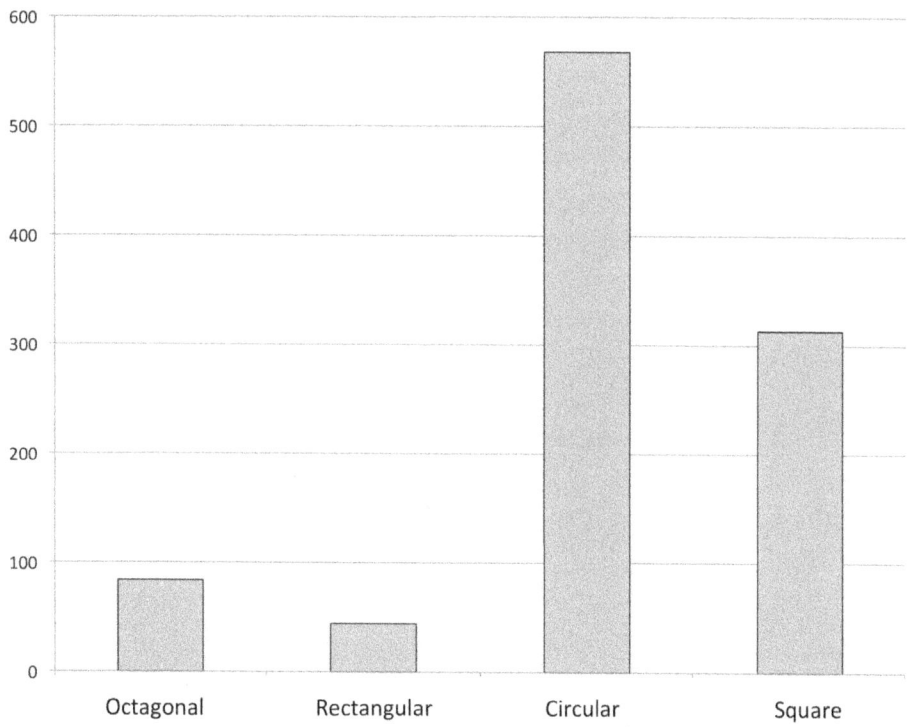

Blockhouses manufactured by 23rd (Field) Company RE from January 1901 to May 1902: 1,009 blockhouses in 17 months, averaging 60 per month.[161]

In Middelburg, Major Rice established the first factory with 23rd (Field) Company RE. Even for the production of the first 50 octagonal variants, a production line was in place manufacturing templated parts. Each station on the line manufactured the same piece repeatedly with the same men to ensure consistency. The resulting blockhouses fitted together well, unlike some that were being produced by contractors on the coast. When required, the factories also took back existing blockhouses that required repair when the lines were moved, and undertook refurbishment tasks. During their stay in Middelburg the company produced over 1,000 corrugated iron blockhouses for use in the district or which were dispatched to other areas.

CRE Pretoria's blockhouse construction yard;[162] notice the A-frame for the roof and already curved walls to the right of the picture. The blockhouse factory took up a considerable space and also had a rail line for onward delivery of the kits. Courtesy of The National Archives, London.

BLOCKHOUSE FACTORY UNITS

In Bloemfontein, 20th (Fortress) Company RE and the remnants of the 7th (Field) Company RE Headquarters were employed in the RE Workshop established to make blockhouses and their accessories. They also assisted in the erection of barracks on Naval Hill overlooking Bloemfontein town[163] and would have been responsible for the Rice Pattern blockhouse originally built there and now at the War Museum in Bloemfontein. Lieutenant-Colonel PT Buston RE and Major JSD Purvis of the RE District Staff used civil labour under command of the engineers in place of military labour, as opposed to contractors whose standard of work was always variable. Any contracted blockhouses were first erected to ensure they were fit for purpose before they were deployed for an operational line. In this area all new blockhouses produced went for new lines, and old ones came back to Bloemfontein for refurbishment.

Reserves of prefabricated blockhouse kits were kept at all manufacturing stations, which, for example, allowed for the Vryburg to Mafeking line in the Lichtenburg District to be constructed from over 400 blockhouses in the reserve in only six weeks during March 1902. Blockhouses were sent from Middelburg, Standerton, Pretoria, Elandsfontein, Bloemfontein and Cape Town to complete this line.[164]

In Pretoria, Lieutenant-Colonel RC Maxwell RE commanding the 47th (Field) Company RE set up a factory where all the blockhouses made by contractors were first reassembled before being transported to the field, as part of a quality control process. Thirteen kilometres east of Johannesburg, at the strategic railway junction at Elandsfontein (now Germiston), the Royal Engineers set up another factory to supply blockhouses. In Standerton a workshop and blockhouse factory was started under the 17th (Field) Company RE for their area. Kimberley District was also producing blockhouses as Lieutenant-Colonel CD Learoyd CRE Kimberley District notes: in the first quarter of 1902, 147 were made, at a rate of approximately 50 per month, just over one-and-a-half per day![165]

EXPANDING CROSS-COUNTRY

Blockhouses were either built by a small contingent typically of nine RE officers and 30 black labourers according to the *Royal Engineers Journal* of January 1902, or regiments were tasked to lay their own blockhouse lines under the supervision of RE officers and the labour force was often supplemented with black labourers. These construction parties needed protection and often columns roving for Boers had to be diverted to this duty. Mobile blockhouses were also used as the lines stretched out across the land, so the movable blockhouses were used to protect the unfinished lines. As the lines moved across the veld, once a blockhouse was completed it was manned immediately and the garrison carried on from the initial build, to complete all the other defensive elements and make living conditions as good as possible. It was the construction party at the head of the blockhouse line that was most vulnerable and likely to be attacked, as shown in the deployment diagram on page 136. It was on one such occasion that a construction party was attacked at Groenkop on Christmas Eve in 1901 by the notorious Christiaan de Wet, tempted by the supply wagons laden with food and ammunition.

From the various railheads, ox-wagons were employed to ferry the blockhouse construction kits and the black workforce to site where building quickly commenced. The construction team for each blockhouse was an RE supervisor, about 20 infantrymen and numerous black labourers. Demand for the RE cap badge supervisor was such that on occasion the Company's Tailor had to become the Superintendent of Works!

A blockhouse in kit form travelling to site.[166]

The construction of the cross-country lines was certainly more of a challenge, there being no convenient rail line from which to drop off blockhouse kits and a work force every 1,000 yards (900 metres). Now the only way to transport blockhouses was by wagon from a convenient stores dump at a railhead. The two methods were either by mule wagon or ox wagon, as follows. The ox-wagon[167] of 16 oxen with two African drivers, as used in the Anglo-Zulu War, could carry up to 6,000 pounds (2,720 kilograms) and seldom exceeded 12 miles (19 kilometres) per day. The main advantage was that they could feed themselves off the veld; the disadvantage was that they could only work for six hours a day, eight hours being required to graze, eight hours to chew and two hours for rest! Compared to the mule wagons they were a heavy-duty approach. Mule wagons[168] with a team of 10 mules and two drivers could carry up to 3,500 pounds (1,580 kilograms) for a maximum of 18 miles (29 kilometres) per day, and required fodder, which was part of their load. One blockhouse could be carried on a mule wagon and might be accompanied by an escort of up to 30 mounted troops. Passage for all was hard going across country; especially when there was water to cross or muddy conditions.

There are claims that six men could erect this well-designed prefabricated kit in six hours, however the reality as reported by Bethell after the war cites a more realistic timeframe. He reported that the construction of a single blockhouse[169] by '3 Sappers, 20 Infantrymen, and black labourers took 6–9 hours. This is most likely only for the basic construction, assembly of the walls, loopholes, the walls filled and roof on, but is still 172 man hours. One of Major Rice's subalterns, Lieutenant C Mellon RE, states that 'Under favourable conditions, one blockhouse can be erected in 4½ hours with 5 Sappers, including barbed wire entanglements, a circular trench and parapet, and "Bird Cage" wire crinoline.'[170] He fails, however, to state how many men are used to accomplish this feat! His normal contingent would have been nine Sappers and 30 black labourers; if they were all employed to build one blockhouse this would equate to 174 man hours to do the same task, quite a different story! The article goes on to state that the transported weight of a blockhouse on a single ox-wagon is 3,500 pounds (1,500 kilograms), but fails to state what is included in the materials. The walls alone would have contained approximately 4 to 8 tons of shingle, the revetment another 30 tons of stone, the sentry trench approximately 90 tons of soil, before starting on the interconnecting trench and wiring. Practice and experience certainly shortened the construction time, which was initially measured in days.

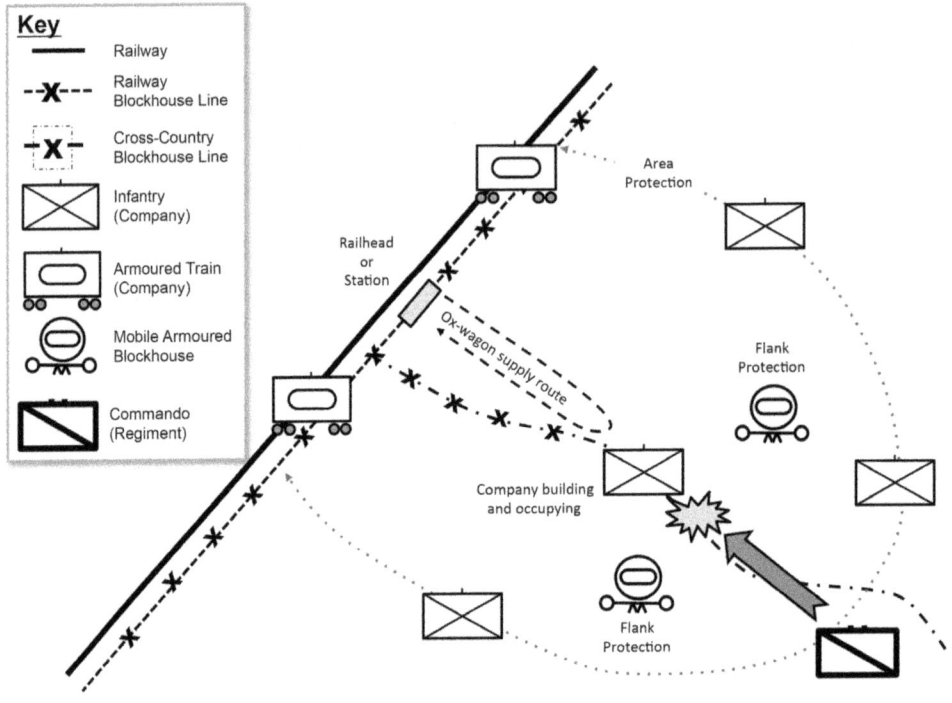

Cross-country blockhouse line construction deployment.[171]

ONE BATTALION'S BLOCKHOUSE BUILDING EXPERIENCE

A first-hand account from the York & Lancaster Regiment in the area of Amelia states that they were assisted by Captain von Hugel RE, who gave all building instructions to the working parties. They managed, with some black labour support, to erect four blockhouses per day, and distances apart varied from 500 to 3,000 yards. The distances varied according to the nature of the ground and care was taken to cover any 'dead-ground' through which the Boers could pass. This line was constructed using both the traditional high-gabled Rice Pattern blockhouse and the lower flat-roof Pepper-pot Pattern blockhouses. The working party numbers decreased daily as the occupying garrisons were left behind, with officers stationed at strong posts and other halfway or intermediate blockhouses. These command posts were most likely company and platoon bases. Work was hard and lasted from six in the morning to six at night, only stopping briefly for dinner of biscuits and tinned rations. One blockhouse was built at Hout Nek (later called Witkoppies Post). Working 12 hours per day, they continued until pausing at Rooi Nek for Christmas Day 1901, on which day they only constructed two blockhouses. The regiment continued to build and fight on and like very many units spent their third New Year's Day in South Africa. They had built a line of 68 blockhouses from Botha's Pass southwest across the veld for 45 miles (72 kilometres).[172]

Life in general for the battalion was described as being monotonous; the men were happy to receive any reading material and would hike over three miles (5 kilometres) to obtain one old newspaper! There would be times of extreme pressure or of action often preceded by intelligence or the rumour of a Boer breakthrough. The Boers would snipe at the blockhouses from long range, forcing the occupants to hunker down to protect themselves. Attacks and breakthroughs would come mostly at night. During such engagements blockhouses could fire 400–2,000 rounds and ammunition husbandry became important in these protracted firefights. Blockhouses would wake to find their fortresses peppered with Boer rounds; in one attack either a lucky shot or a fine sniper managed to hit a soldier's rifle through a loophole as he was returning fire.

THE WHOLE SYSTEM DEPLOYED

The blockhouse system that existed at the end of hostilities in May 1902 was the largest group of fortifications ever built in South Africa. The Royal Engineers' summary return[173] dated 12 May 1902 records a total of 441 masonry, 7,447 corrugated iron blockhouses and 555 works built under their jurisdiction, to which must be added an unknown number of infantry and Town Guard forts, all distributed throughout the area of operation. Detailed as they may appear from original records, the established and accepted figures '3,700 miles of lines and up to 8,000 blockhouses'[174] do not appear to be accurate. Extensive transcribing of the records from all sources, including Bethell's Report, official Boer War records, original site maps of blockhouse lines from the archives and author's collection indicates the conservative figures are closer to nearly 6,000 miles (9,600km) of blockhouse lines comprising nearly 9,000 blockhouses. Even the Royal Engineer records from theatre, from which most observers of the period have taken their figures, have mathematical errors! The sub-total of 8,018 blockhouses is actually 1,640 over the listed quantities per district, which adds up to only 6,378 corrugated iron blockhouses.

District	Blockhouses (Brick)	Blockhouses	Work
Barberton	10	98	21
Middelburg	34	82	34
Pretoria	58	196	–
O.C. Life N.		117	4
Germiston	20	71	64
O.C.W. of Johannesburg	43	488	30
Heidelberg Middelburg	50	270	106
Natal	15	538	52
O.R.C.	58	1013	189
Harrismith	12	467	3
Aliwal North		333	6
C.C.D.			
Midland		449	
Eastern	79	311	
Western	9	524	14
S. Western	25	230	
Southern		26	8
N Western		89	
Namaqualand		45	
Kimberley	17	352	
Western District	11	570	4
General Total	441	8018	555
deduct those added in April & May		571	
Total Blockhouses & Posts 31/5/02	441	7447	555
		7888	555
		8443	

Table of blockhouses and posts built up to the end of the war.[175]
Reproduced by kind permission of The Royal Engineers Museum, Library & Archive.

The details of individual lines are meticulously recorded in Appendix A
Key assumptions used are as follows:
- Where the actual number of blockhouses per line is not recorded, a conservative average blockhouse per mile is used.
- Where distances are not given these are measured accurately from maps.
- For the rail lines to Port Elizabeth and Port Alfred, no defences were included south of Queenstown.
- For the rail line in Natal no defences were included south of Dundee.
- No Town Guard forts are included in the total, which it is estimated could be another 240–500 dry stone forts.
- Where figures are available town defences are included.
- South African Constabulary (SAC) posts are included.

- The total does include lines that were moved, for up to 144 miles' (230 kilometres) worth of lines, amounting to approximately 200 blockhouses.[176]

The following table is a breakdown of the lines in each category and the interval between each blockhouse from actual and average distance given from all historical sources (again these are conservative where estimates are made).

Line Type	Miles	BHs	BH/mile	BH built per
Rail	2,951	4,876	0.6	1,000 yards
Cross-Country	1,851	2,878	0.6	1,000 yards
River	397	458	0.9	1,500 yards
SAC	698	719	1.0	1 mile
TOTALS	**5,897**	**8,931**		

Author's calculated blockhouse line distances.

Many who arrived in theatre later in the campaign, such as Colonel Ian Hamilton, who joined Kitchener's staff, remarked on the sheer size of the operation. Hamilton remarked in November 1901 that: 'Although I had read much of blockhouses, I never could have imagined such a gigantic system of fortifications, barriers, traps and garrisons as actually exist. This forms the principal characteristic of present operations, supplying them with a solid backbone and involuntary loss of territory to the enemy which former operations did not. Thus certain areas are permanently under our control. Subsidiary areas are consequently now giving less and less trouble.'[177]

That however is not the whole story. Blockhouses we can easily label and categorise, but there were other fortifications called 'works' which get little mention and are most likely much less permanent. In addition there were the South African Constabulary posts, which were most likely blockhouses or larger forts, but with few records or photographic evidence. In the Cape Colony there were also very many towns to protect, by the formation of Town Guards and their supporting forts, and older trade routes through the mountains. Geographically these passes, poorts and kloofs were also key 'choke points' that needed fortifications, many of which still remain due to their remote locations.

PASSES (POORT, NEK AND KLOOF)

Gradually throughout the war the strategic assets that were required to fight the war and control the population in the former republics were coming under British control. The railways as lines of communication had been protected and fortified with masonry and corrugated-iron blockhouses, the three key towns of Bloemfontein, Johannesburg and Pretoria had been encircled with blockhouse lines, and areas of veld were further contained with cross-country lines. The only remaining key strategic choke points were geographical, and the land had several large mountain ranges, with key passes that also required control and denial to the Boer commandos. Often in very remote areas, these passes (poort, nek or kloof in Afrikaans) were also controlled through the use of small forts or blockhouses.

There were three major mountain ranges that required to be sealed off and denied to the commandos: in the Cape region the Cederberg Range close to Cape Town's north western approach, and the Langeberg Range in the Klein Karoo that largely protected access to Port Elizabeth; and the Magaliesberg Range to the west of the approaches to Pretoria and Johannesburg. Those areas close to towns or railheads could be supplied with either stone and mortared buildings such as in Kommando Nek west of Pretoria, or with Rice Pattern blockhouses from one of the local factories. More remote areas in the Cape and Klein Karoo passes usually had more ad hoc dry stone structures built by the town guard using local stone from the mountains, and were usually termed forts and not blockhouses.

SAC POSTS

The South African Constabulary (SAC) was mostly used in the Orange Free State and Transvaal Republic to control areas of land already occupied by the British Army, later in the war. They were largely raised from Britain during the general drive for more manpower to the total of approximately 10,000. Organised by Baden-Powell, they were designed to form a more permanent British garrison in South Africa and formed a crucial injection of British settlers, part of Lord Milner's plan to anglicise the country.

The SAC termed their fortified sites 'posts' and they could be a variety of large or small garrisons of men strung out in line across the veld but not designed to be strongly interconnected. They were to act more as bases from which to conduct patrols and often had large mounted contingents stationed in them with stables for horses. In this regard they required a large stable or remount area and a perimeter often protected by earthworks, sandbags and blockhouses.

These sites were often spaced quite far apart and were a forerunner to the more closely spaced blockhouse lines. SAC Post constables sought to interact with the local population, gain trust, police local matters and gain valuable intelligence for military operations. Their original role however was to augment the fighting troops and testimony to that fact is that they had many casualties. The monument to the war in Petrusburg, as an example, bears the names of many who died as members of the SAC during the campaign.

As part of this policing force there were several components including the Natal Mounted Police and also a Native Police Force, as shown in this photograph of the period. Both were organised on traditional British colonial lines with a white officer infrastructure and lower ranks of mainly poorly equipped black auxiliaries or constables. The Natal Mounted Police were already in existence at the start of the war, but additional forces were required to manage and police the local black population. As we see they are rather smartly dressed along quasi-British police lines, having a good uniform, traditional police helmet, handcuffs and armed with a knobkerrie or club. Although used to augment the armed effort and release army units to fight, their focus later moved to control and more traditional policing roles.

Men of the Native Police Force c1900.[178]
Reproduced by permission of National Army Museum Archive.

FORTS AND REDOUBTS

The need to position and garrison larger bodies of troops also required the construction of larger fortifications, usually termed 'forts' and 'redoubts', defined as follows:

A fort[179] is defined as 'a work established for the defence of a land or maritime frontier, of an approach to a town, or of a pass or river'.

A redoubt[180] is defined as 'an outwork or fieldwork, square or polygonal in shape without bastion or other flanking defences, sited at a distance from the main fortification, used to guard a pass or to impede the approach of an enemy force'.

They were mostly constructed with earthwork ramparts and ditches and had key points such as entrances and corners protected with sangars or small stone guardhouses, as shown in the photograph of Fort Jackson below. Although many blockhouses are named as forts and redoubts by the troops who built and occupied them, technically this is incorrect. The terms are most often used with the names of the landowners or generals for ease of reference, and perhaps to make an isolated blockhouse feel more like home. Many of the remote forts and redoubts can still be identified and have left a significant mark on the landscape from their perimeter earthworks. A good example is the one for Steinaecker's Horse near Kaapsche Hoop in Mpumalanga, discovered from the distinctive hexagonal shape, recognised by a researcher on Google Earth.

Fort Jackson, a Devonshire Regiment redoubt, with breastworks.
Courtesy of Tim Saunders, The Devonshire Museum.

TOWN GUARD FORTS

The British had used the garrisons, railways and ports of the Cape Colony to launch their attacks against the Boers in the initial stages of the war. At this point there was little concern of a Boer counter-attack, at least until the fateful reversals of 'Black Week' in December 1899. The Cape Colony towns and dorps had now become vulnerable to attack, which prompted the formation of Town Guards, one early one of which was Port Elizabeth in February 1900. Later the invasion of the Colony by General Pieter H Kritzinger's commando in late December 1900 and various other commandos at a later stage prompted the formation of local guards, known as Town Guards. They were a precursor to the Home Guard (so-called 'Dad's Army'), which was formed from local volunteers to protect British towns from invasion during the Second World War (1939–45).[181]

British troops were generally fully committed to the push into the Republics, then in fighting the guerrilla war and in guarding the lines of communication. There were insufficient troops to guard these small towns in the Cape Colony and then later in the occupied territories further north. Local units were formed as volunteer regiments or smaller units to protect local towns and infrastructure sites, such as reservoirs. Particularly in the Cape Colony region, the British were continuously concerned about an uprising of the Cape Dutch population in support of the Boers, but over time this failed to materialise. On 14 November 1900 Milner argued that the situation in the Cape Colony was now becoming critical and that there was a need to raise local town guards and to impose martial law. On 31 December 1900 the Cape government authorised the formation of the Colonial Defence Force (CDF), which was a part-time contingent that might be expected to serve beyond the Colony's borders.[182]

Town Guard contingents were raised in over 128 towns across the Cape Colony region and into the now occupied former republics. If each town was guarded by only 2–4 fortifications this would amount to another 240–500 other smaller fortifications, of dry-stone construction, that are not recorded anywhere. Sites such as Uniondale are typical of these small forts. They were typically designed to be 30–100 strong in the smaller towns. The Cape Town Guard was 4,000[183] men strong, commanded by a captain and four lieutenants. The boss of a commercial company down to the lowest office boy might serve in the contingent. There were several different units formed: mounted troops; a Cyclist Company, a Sailor's Company and an Ambulance Section – one even manned an armoured train.[184] Few of these Town Guard units saw action, but all were committed to local defence of their area and built many small forts, redoubts and blockhouses for the purpose. The attested strength of the whole Town Guard force was 20,140 men.[185]

Kritzinger's commando, a formidable fighting force to be reckoned with despite shortages of food, clothing and ammunition. The towns of the Cape Colony became prime argets for morale-boosting and courageous raids by these Boer units.

As defensive protection many different types and styles of fortifications were built, no doubt depending on the local skills and building materials available at the time. Access routes to the towns and the immediate town itself were guarded with either sandbagged shelters or blockhouses, largely to provide a piquet to control access. Secondly the surrounding hills were controlled by blockhouses made from local stone, such as those at Barberton, Montagu, and Uniondale, where there could be as many as six providing oversight of the town and its approaches. Finally several forts were built to control strategic sites, such as water supplies and access points. Examples of this type of defence remain intact at Port Elizabeth guarding water supplies, Harrismith guarding the town's reservoir, and Cogman's Kloof overlooking the key road and tunnel into Montagu.

Locals also served and fought in many of the town guards, particularly in the Cape region. Many coloureds sought to enlist but were initially deemed 'ineligible', despite there having been coloured regiments in the Cape since the 18th century. There was still a huge political sensitivity to keeping anyone of colour out of the fighting and unarmed, though this did eventually relax. Most notable actions involving coloured volunteers were in the battles in O'Okiep and Springfontein where they fought against the commandos who besieged the towns and mines there later in the war.

Blockhouses were usually of a dry stone construction, and made of locally available or easily quarried stone. Some were constructed from mortared and crenellated stonework, such as Knysna Fort, but few of these remain intact today. One such example is in Montagu in the Eastern Cape where the town recruited the Montagu Town Guard and a Montagu District Mounted Troop to defend the area from Boer incursions. There were nine forts built to protect the town, situated on the surrounding hilltops. These units fought alongside their regular counterparts and were also awarded medals – in this case there were 35 Queen's South Africa medals awarded to the Montagu Town Guard.[186]

SUMMARY: HOW THE BLOCKHOUSE LINES SPREAD ACROSS THE COUNTRY

The four maps on the following pages summarise the deployment of blockhouses from the beginning of the occupation of the Boer Republics in June 1900 until the end of the guerrilla phase in May 1902. They highlight how the strategy based on the need to defeat a highly potent mobile enemy had to be harnessed with the rapid development of new operational concepts and tools. Virtually every part of the operational area had some type of blockhouse: the plethora of types used were all designed with the same purpose, to defend a small area and deny Boer movement.

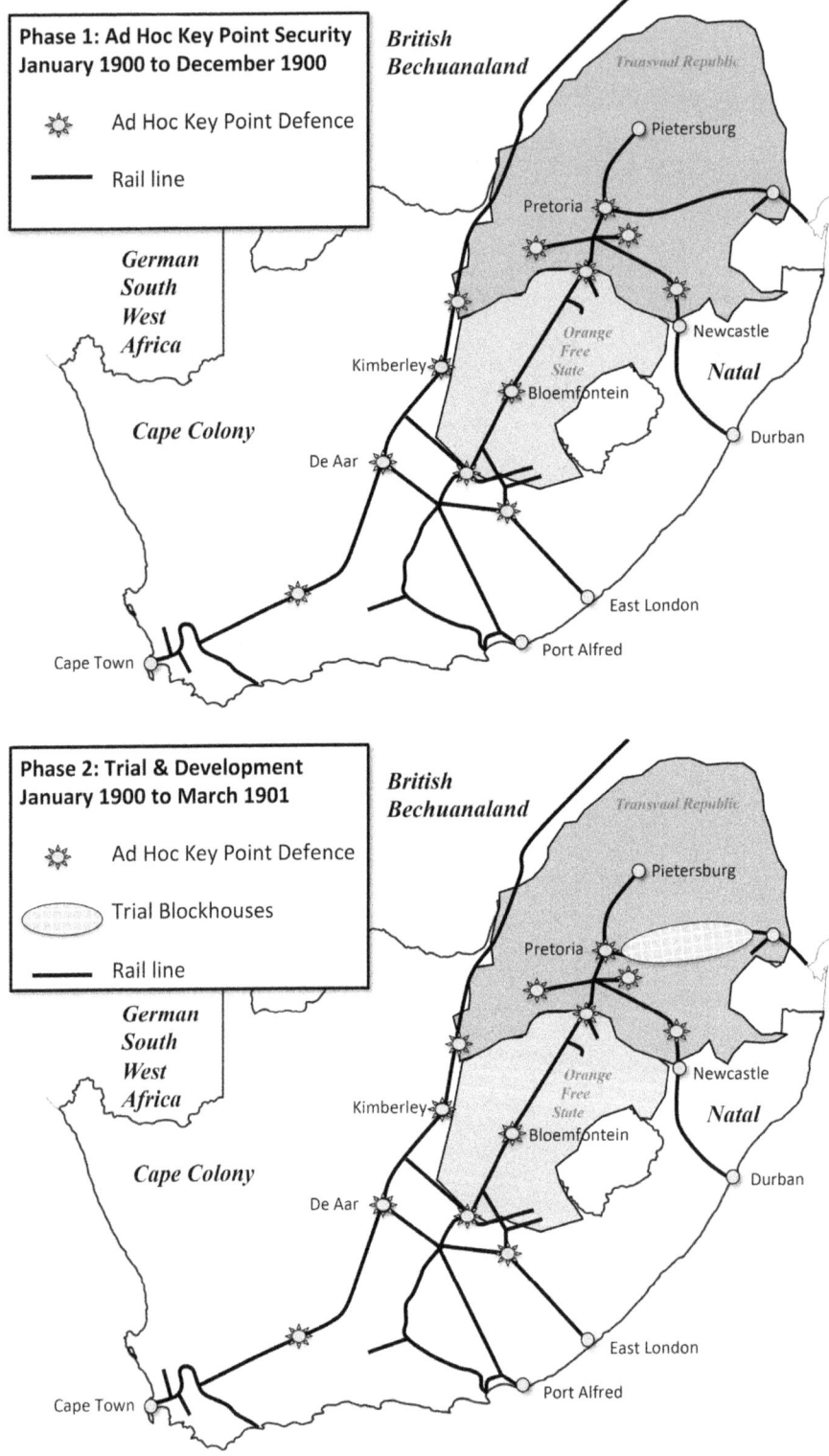

EVOLUTION OF THE BLOCKHOUSE SYSTEM IN SOUTH AFRICA

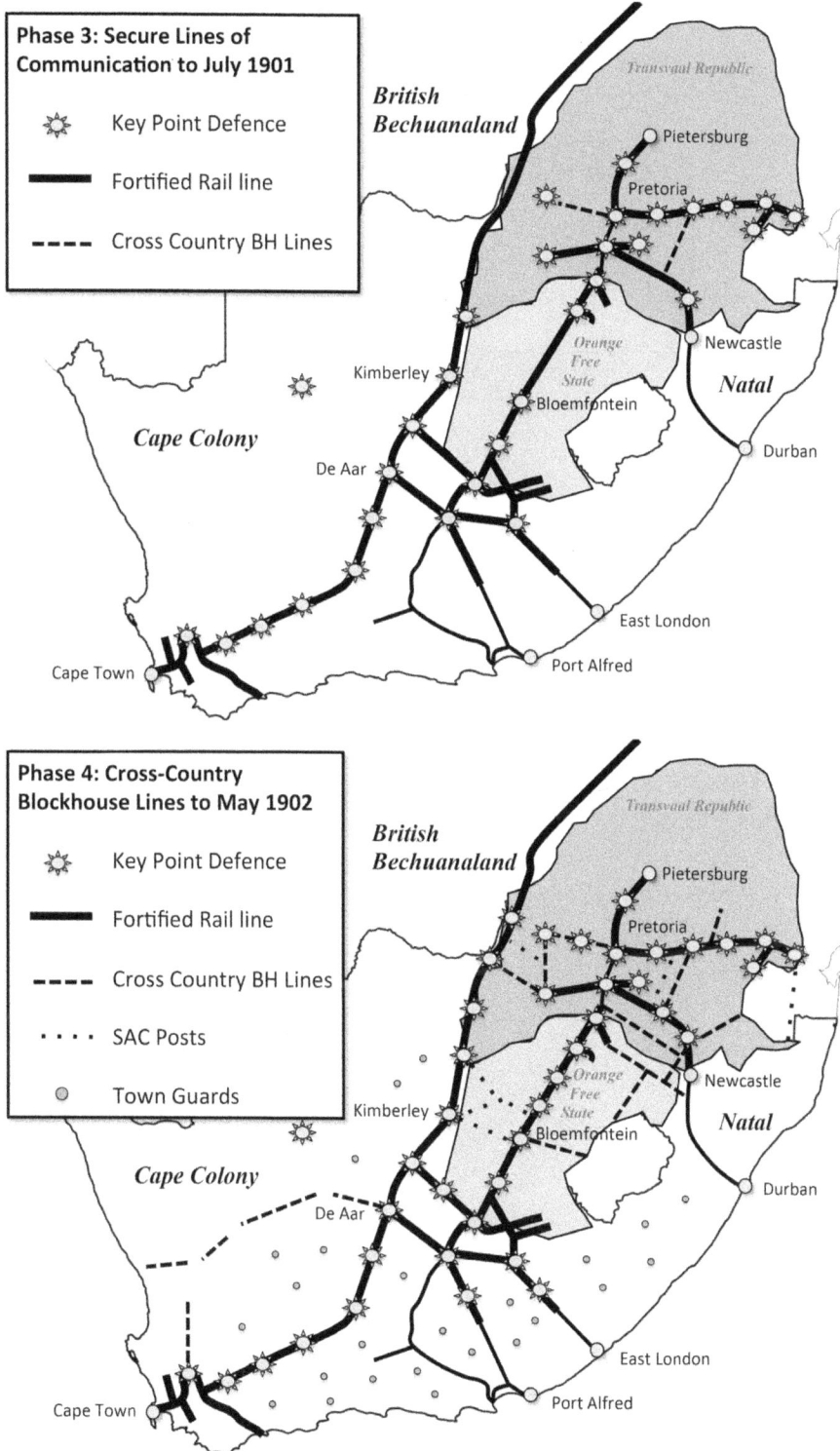

The four phases of blockhouse system deployment.

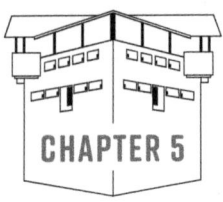

CHAPTER 5

BLOCKHOUSE SYSTEM COMPONENTS

The whole is greater than the sum of its parts.

ARISTOTLE, *METAPHYSICS*

The *Oxford English Dictionary* defines a system as: 'A set of things working together as parts of a mechanism or an interconnecting network; a complex whole.' In this regard and as part of the strategy and tactics developed throughout the campaign the blockhouse system can be broken down into a number of components defined in this chapter as elements to five discrete sub-systems (defined as a self-contained system within a larger system): the human component, protection sub-system; alarm sub-system; communication sub-system and supply sub-system. All interact to form a connectedness that overall is the blockhouse system as deployed in South Africa from 1900 to 1902.

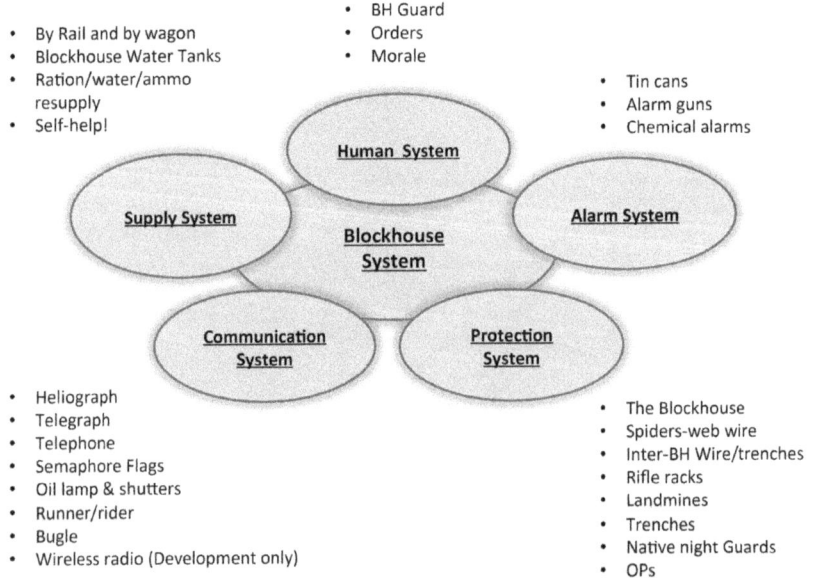

A summary of the blockhouse system components.

HUMAN COMPONENT

British and Imperial Forces were the mainstay of the guard force to man the blockhouse line, although Africans were extensively used to augment this complement. As the lines grew and more were added towards the end of the war, some blockhouses were entirely manned with armed Africans, something the British had said they were not going to do. Virtually every regiment of the line which served in South Africa were either involved in constructing blockhouse lines or found they were garrisoning the line, often for many months at a time. It was only the cavalry, mounted troop or yeomanry units that did not get their stint in the blockhouses due to being needed for mobile columns and drives. Each blockhouse formed a mini-garrison and was generally organised within the regiment's structure as follows:

For a typical company of approximately 230 men a Captain's Headquarters blockhouse was commanded by one officer of subaltern rank and comprised approximately 20 men, and was either a large corrugated iron or masonry structure. The captain commanded his sub-units (four troops or platoons) each commanded by a junior officer who was garrisoned in 10–12 blockhouses for the company.[187] These numbers varied considerably depending on whether the unit was manned to its full war establishment or something much lesser. The burden being placed on the infantry to man and sustain these ever-lengthening blockhouse lines was summarised in a letter by Colonel Ian Hamilton as follows: 'The strain on the infantry is very great at present. Blockhouses used to be held by 10 men, then by 7, then by 6 and now they are cut down to 4 (on certain parts of the line that is to say), which is very wearing, in the matter of sentry duty, and of course does not admit patrols. I must say I do wish most awfully that we had another 10,000 infantry of any sort. It would make matters simpler.'[188]

At the same time many men were also being drafted out of foot battalions to furnish the numbers required for additional mounted infantry companies to match and counter the mobility of the Boer commandos. On the skills required of the men left in the blockhouses Hamilton goes on to remark: 'We only want a shooting machine in a BH of course we prefer a highly trained, vigilant individual but a shooting machine who is not a coward will do all right.'[189] Fuller, who had served as a junior officer during the war, later commented in his book that the work was 'demoralising, and that the men could only talk, smoke and gamble. There was a monotony of food, and no civilization for miles around, it was like living in a medieval castle for days on end ... masses of men were bored to extinction.'[190] Morale was not high and the officers and non-commissioned officers faced a daily battle to keep the men occupied and alert.

Black troops were used in a variety of roles, such as on the Imperial Military Railway system, and served as scouts, labourers and wagon drivers, in addition to extensively serving on the blockhouse lines. They also became refugees of the war, had their homes and livelihoods destroyed in the British scorched earth policy, and were interned in the notorious British concentration camps, but for blacks only.

Although both the British and Boers initially agreed that black people were not to be used in a combatant role in the war, official sources state at least 15,000 blacks were armed by the British and served in mobile columns to track down Boer commandos. A further 25,000 at least served as armed blockhouse guards, although Kitchener always sought to cover this up to the British government.

A Section of seven men of the Canadian Mounted Rifles (with pets) 1902.[191]

In July 1901 General PH Kritzinger in the Cape Colony had warned Lord Kitchener that if 'blacks or coloureds employed by the British were caught by the Boers, they would be executed',[192] whether or not they were armed. As a direct result of that threat, both the British Army and government decided to arm even their black scouts for self-defence. Hence, from the latter part of 1901, most black scouts attached to British forces in the Transvaal and the Orange Free State were given arms. Following this concession, other categories – for example, those guarding blockhouses and blockhouse lines – were also issued with arms. In response to severe criticism in Britain of this practice, especially by the Liberals, General Kitchener was forced to disclose that in both Boer Republics combined there were 'over 10,053 armed blacks under British Army control'.[193]

For blockhouse duty these men might have had a tent outside the main blockhouse but inside the inner perimeter, and were used for local intelligence, night-guarding, manning observation posts, and for general duties building and maintaining defences. For this duty they were variously equipped with clothing, rationed and given arms, no doubt being provisioned by the blockhouse guard unit.

The Army of the day was very much a microcosm of the society from which it was drawn, its officers and men drawn in this case from the well-educated upper class for its officers, and the working classes for the rank and file. Rank and file soldiers of the day were described, perhaps a little harshly, as 'indifferent shots, slow to comprehend what was taking place or to grasp the whereabouts of the enemy, always getting surprised or lost, helpless without their officers'.[194] Typically up to that point structures and tactics had relied on strong mass unit formations led for the most part in fluid battle by their officers and senior warrant officers.

Throughout the Victorian era the military was transforming and went through two reforms, with the Cardwell Reforms of the 1870s and the Childers reforms of 1881. Many of the changes were prompted by the Army's performance in the Crimean War in 1854, and also by more exposure to reporting in the press and avid readership in *The Times* newspaper. Throughout the 19th century the Army had been involved in overseas operations on behalf of the British Empire and the opposing forces had often been lesser-equipped 'native forces'. Technology stemming from the industrial revolution had led to the development of better weapons such as rifles and artillery, and more general technology such as camouflage, radio, telegraph and railways had been fully incorporated and utilised by the British army entering the Boer War. Old traditions and interests die hard, however, and it was not until later in the period that Liberal governments managed to make broad changes that the Army retained until the start of the First World War.

Terms and conditions of soldiers had generally increased throughout the century, and in general they were better paid, better fed and accommodated and were better educated than their predecessors. Soldiers now wore uniforms of olive drab or khaki – which originates from the Urdu word for dust and from the uniform of that colour worn at the time in India: gone were the scarlet tunics of previous Boer encounters from 1881; the Scottish regiments even had olive drab covers for their tartan kilts. The infantry weapons had moved from muskets fired in volleys to breech-loading rifles, and there had been the introduction of rifle regiments specialising in skirmishing and 'sharpshooting', such as the King's Royal Rifle Corps. Additionally rapid-fire or machine guns serviced by a small team were introduced, such as the Maxim, and proved a formidable force multiplier in firepower. Soldiers in the regulars usually served for between four and 21 years, though it was a minority who passed through the ranks to senior non-commissioned officer and warrant officer levels. There was also now an established reserve and the early leavers then continued some sort of reserve service.

Infantry and Cavalry officers no longer purchased their commissions and progression was now by a system of advancement by seniority and merit, regardless of personal means. Although commissions were now obtained by attending a military academy, most cadets had to pay their own way in terms of education, uniforms and other equipment. Naturally therefore officers were still drawn from the classes that could afford this expense, largely coming from the aristocracy or landed gentry. There was an expectation of an expensive lifestyle and officers still required a second income to supplement their officer's pay. Army staff work was beginning to come to the fore as the military's campaigns and warfare became ever more complex. Unfortunately although there was a new staff college established in Camberley, unit quotas and favouritism persisted to deny some capable officers their opportunity to attend and progress into the general staff. In addition to this and as a result of much family and regimental nepotism, the officer classes' ability to lead complex and demanding operations was somewhat inconsistent at best. At the senior level the generals were equally competitive for promotion and formed their own circles or 'rings' of officers they recruited and favoured. General Garnet Wolseley, based on his operational experience and background, had established the 'African Ring' and General Frederick Roberts the 'Indian

Ring'. This resulted in quarrels between the factions and disputes in the appointment of officers to what were seen as prestigious appointments, a situation that was not resolved until most of the officers had retired from the army.

PROTECTION SYSTEM

The Blockhouse

Much like the 'Motte and Bailey' or 'Castle and Keep' concept, the blockhouse itself stood at the centre of the protective system of defences. It was both the soldier's place of sanctuary and defence from the enemy and his home from the harsh elements of the South African veld, the sun and heat, rain and the cold of the winter's night. It could be either a large low-profile blockhouse styled on a bunker with the majority of the shelter below ground, or a smaller but raised three-storey masonry blockhouse designed to stand as a sentinel to deter attack at strategic sites. The various types have already been described in terms of size and location. As with any defensive system it was largely fixed, although a small number of corrugated-iron blockhouses lines were moved, and designed to send an overt message to 'stay away'. Once the initial construction was completed it was garrisoned by the established number of men, and then the work really began. Often it would take weeks of hard toil under the sun to complete the total defensive system, starting with the sentry trenches.

Each blockhouse was sited according to the tactical need and, where digging permitted, was to be provided with a circular trench for sentry duties. The type of ground usually dictated the depth of the trench and the time that it took to dig. It was about 4 yards (3.7 metres) from the blockhouse and 4 feet 6 inches (1.4 metres) deep and wide, and was constructed to provide cover from view and protection for the sentry. Fire could be brought to bear along the fence lines, with troops coming out of the blockhouse at night to fire down the fence line, which was at an angle to the blockhouse line. Access to the rifle racks down each fence line was also possible and allowed them to be reloaded once initially fired. The sentry would have been the night-time eyes and ears of the men sleeping inside, initiating alarm gun and rockets if attacked. The African night-guards were also used, and observation posts set up directly between the blockhouses so as not to be in the line of fire down the fence line.

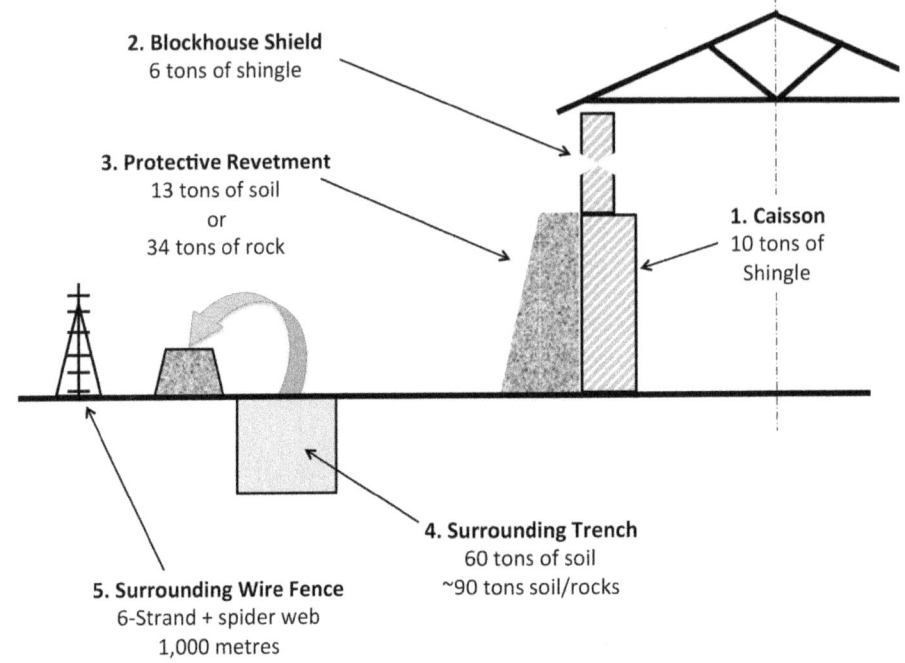

Estimated quantities of material used in defensive tasks.

Fixed Rifle Racks

The blockhouses covered the fence and ditch obstacles and once alarms were raised in the dark, fire had to be effectively brought to bear along these lines. When they were within rifle range of each other, they were set out specifically in a zig-zag pattern so as to avoid 'blue-on-blue' fire between blockhouses. Fire could now be directed along the fence lines without fear of killing men in the next blockhouse or the black night-guard if deployed.

Each Rice Pattern blockhouse had 12 loopholes so from each blockhouse only two rifles could be brought to bear down a single fence line from inside; the others were rendered useless for this effect. Also the two rifles firing through the small loopholes at night with no illumination made the fire ineffective. Fixed rifles, mounted on wooden racks such as that in the photographs below were preferred, but depended on the availability of spare rifles. It was intended that each barbed wire line extending from a blockhouse would have two rifle racks sited on either side of the wire, suggesting that four racks per blockhouse might be used.[195] The racks could hold 2–6 rifles and were command wire-fired from within the blockhouse. As the rifles used were bolt action, this only provided a 'one-shot' capability, as the rifles had to be manually reloaded once fired.

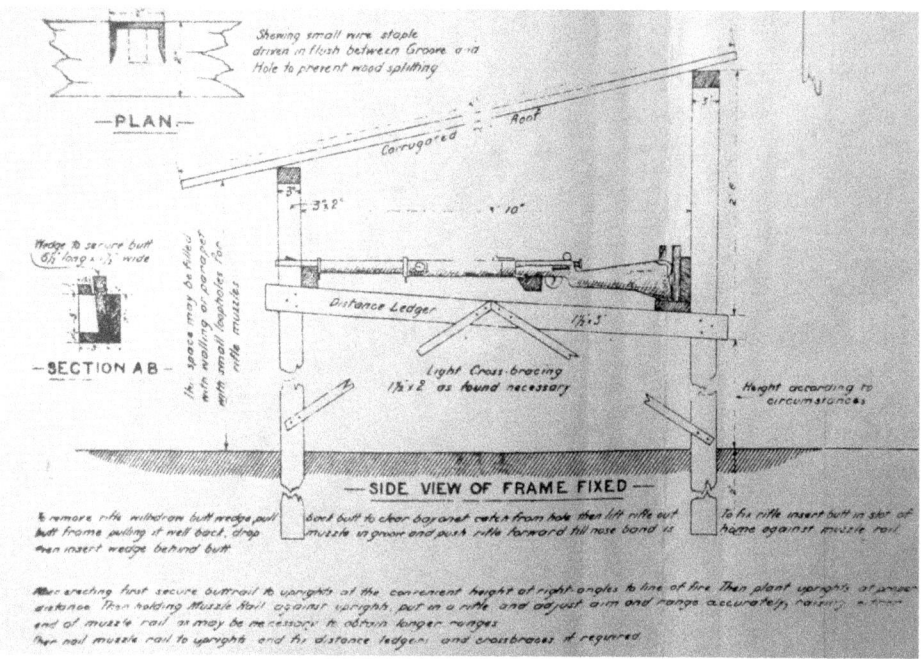

Post-war documentation showing the design of the rifle racks, usually containing 3–4 rifles.[196]

From each end of the wire, then, a maximum of 12 rifles fired a single volley of fixed-fire, otherwise the individual rifle fire relied on sound and firing in the general direction of the Boer breach. It was suggested that these fixed rifle racks could then 'sweep' the line and effectively provide fire at night against Boers attempting to cross the line. However there is little photographic evidence to suggest that this scale of rifle rack deployment was ever achieved, or if these were additional rifles to those already held by each soldier inside the blockhouse.

Rare photograph showing the deployment of fixed rifle racks on the Orange River near Aliwal North.[197] Photograph courtesy of the War Museum, Bloemfontein.

Machine Guns

Panoramic views over the vast wide-open veld from the rooftops of large masonry blockhouses made these ideal positions for machine guns. A relatively new technology, rapid-fire machine guns, such as the Maxim-Nordenfelt (or Maxim) gun, was ideal for this type of terrain. Blockhouses were often sited with dominating high ground to observe out to long distance, both to watch for the enemy and for line-of-sight heliographic signalling. Those with closed-in top floors were not well suited for machine guns as they required good all-round observations best suited to provide fire in any direction. Many of the larger open-roofed blockhouses served as strongpoints and command posts, and were ideal for mounting machine guns on. The photograph here shows a Maxim gun mounted on a roof-top gantry that could provide supporting or defensive fire to all directions.

Maxim gun and lookout on Leeuwspoort blockhouse.[198] This was a large hexagonal two-storey blockhouse capable of housing approximately 20 men and commanded by an officer. Courtesy of King's College London Archives.

Ripples of fire down the line

One incident in the night was recounted by the Yorkshire and Lancaster Regiment as follows: 'We husbanded our ammunition in during the night, expecting the enemy to come on at daybreak; but instead of that they retired just before dawn to Witkopjies. The firing from the trenches round the hill kept up all night; one blockhouse fired as many as 2,000 rounds another 450 rounds. Luckily we had no casualties, though some of the blockhouses were very much peppered, and one of the Boer's bullets went through a loop hole, striking the rifle that a man [was] in the act of firing off.'[199]

Protection Fences

Outside the sentry trench at a further distance of 15–25 yards (14–22 metres) a six-stranded wire fence was constructed in order to prevent anyone approaching the blockhouse to attack it. This in addition to the spider's web that formed the close protection element denying the Boers the chance to fire through the loopholes. Again this was a considerable defence task, to dig in the approximately 35 stakes, robust enough to take the weight and tension of the six strands of wire, each about 160 yards (150 metres) long. Encasing the blockhouse in a defensive mesh took approximately half a mile (1 kilometre) of barbed wire.

> ### The History of Barbed Wire
>
> Barbed wire, also called barb wire or 'bob wire' is formed by twisting wire with sharp edges arranged at intervals along it, first developed in the United States of America for cattle fencing. The first patent in the USA was issued in 1867 to Lucian B Smith from Ohio. This type of fencing was cheap to produce and easy to erect by unskilled workers and ideal for the rapidly expanding cattle industry. Based on this early patent and design many other different styles and patterns were produced by different manufacturers. In the late 1870s, John Warne Gates of Illinois began to promote barbed wire, now a proven product, in the lucrative markets of Texas. At first, Texans were hesitant, as they feared that cattle might be harmed, or that the North was somehow trying to make profits from the South. Barbed wire was first used by the military in a large conflict in the Spanish-American Wars (1898) during the Siege of Santiago by Spanish defenders.

Connecting Fences

These started as simple light fences to connect the blockhouses together that could be quickly breached by the Boers at night. Aprons of wire were then added to strengthen the line by adding sloping wires, to which three to four more horizontal wires were fixed. The Boers were still getting through by both cutting and pulling up the stays to overturn the fence line enough to pass over. To prevent wire-cutting, twisted strands of barbed wire were made up to form a thick rope of wires. An annealed steel wire was also used, which at ¼ inch (6 millimetres) thick was 'uncuttable' with conventional wire cutters, but could be severed with a blow from an axe. Stays now had to be more robustly anchored, which meant more work digging them in with stones and often sandbags. For a typical fence there were over 400 stays and anchor points per mile (1.6 kilometres) of fence, and the reinforcing would have been a time-consuming, arduous and gradual process. Improved and hardened wire was ordered from England in August 1901 but the first consignment came late in the war, only arriving in Durban in early 1902.[200]

A Devonshire Regiment party on wiring duties connecting blockhouses together. Photography courtesy of Tim Saunders, The Devonshire Museum.

Anti-Mobility Ditches

Some Boer commandos were supported by light carts and these also had to pass through the lines, often with herds of oxen. To counter this passage, shallow ditches were initially completed, about 3 feet wide and deep (1 metre), which were later made 5 feet wide (1.5 metres) by 4 feet deep (1.2 metres). Later these were reinforced to a pair of ditches of the same depth with the space outside them filled with a small berm of spoil from the ditches. It is difficult to imagine the boredom noted by the men and many observers when there was so much defensive work to be completed (unless the African contingent did it all?) Throughout the final 12 months of the war the blockhouse lines became steadily harder to cross as they were developed and reinforced, although determined commandos did attack and capture blockhouses and even had time to fill ditches in order to cross over.

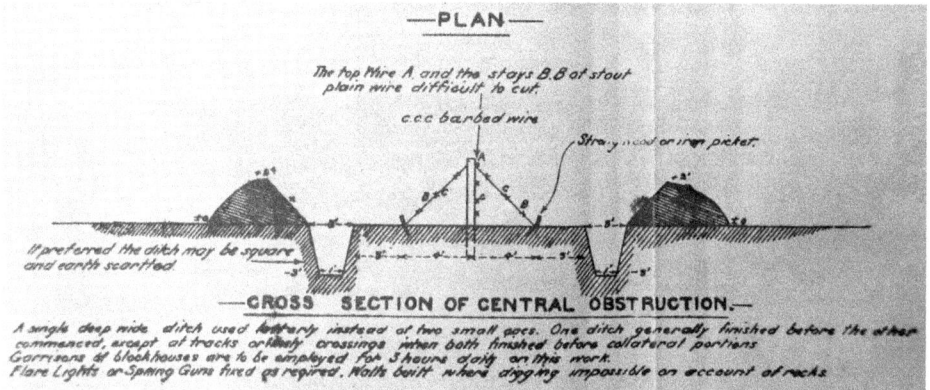

Post-war description of the inter-blockhouse fence and ditch plan.[201]
Courtesy of the War Museum, Bloemfontein.

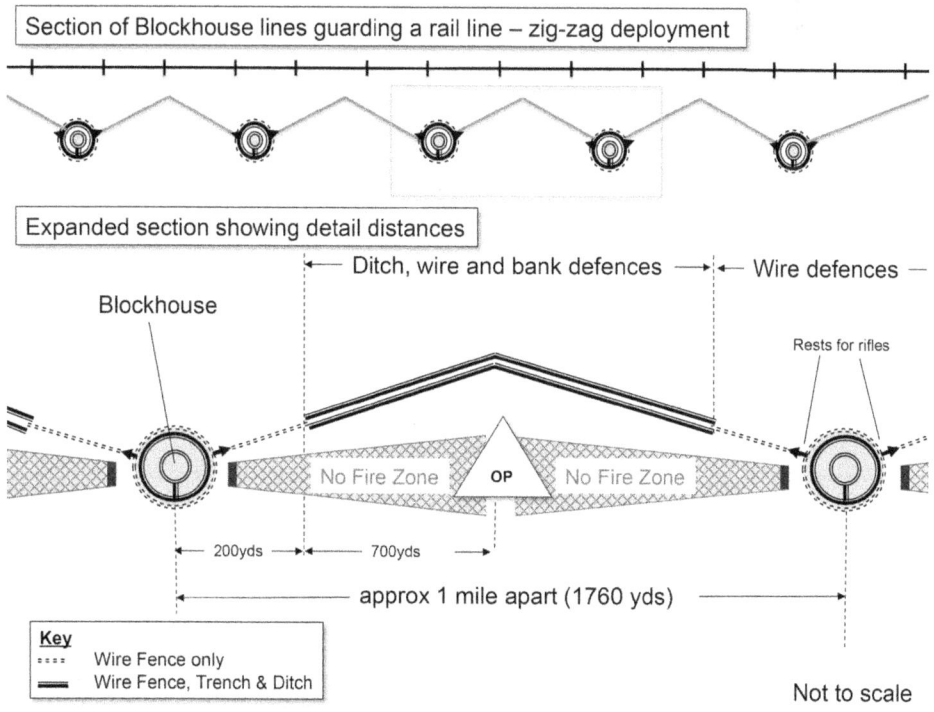

A typical ziz-zag blockhouse line deployment.

THE ALARM SYSTEM

Alarm Guns

Now that the fence-and-ditch obstacle was completed and was covered by fire from the blockhouse, this now required initiation. During the day obviously this could be achieved by observation and orders to open fire; however the Boers rarely attempted to cross during the day, preferring the cover of darkness. Alarm guns (called Spring guns in theatre) were designed to be fitted on the fence posts close to the blockhouse, and set to be fired on a trip wire connected to the fence. When the fence was cut, touched or tampered with, the gun discharged; alternatively it could be command-wire initiated from the blockhouse. The design of these guns was attributed to Major BE Morony RE and several thousand were produced to protect the fences.[202] Later alarm guns were placed at the blockhouse only, as the Boers were spotting them and disconnecting them before crossing.

Total Spring Guns Fitted	By date
140	1 Jan 1902
296	3 Jan 1902
524	6 Jan 1902
600	13 Jan 1902
712	17 Jan 1902

Alarm gun deployment schedule De Aar to Beaufort West, January 1902.

Alarm gun fitted to a blockhouse.[203] Courtesy of SANDF Archives.

Chemical Alarms

Chemical alarms were also used where the breaking of a glass phial combined two chemicals to cause a detonation or to light a flare. They were similar to home-grown types using percussion caps and nails as strikers to do the same. Here boredom was working in favour of the guards, and no doubt many ingenious methods not recorded were also used to create ad hoc alarms.

Electrical Alarms

There are Boer recollections from POW Camps in Ceylon[204] that the system used there by way of fine electrical cables running along the extent of the barbed wire fences, was similar to a type used to warn blockhouses of interference in their fences. The idea is that the wires are so fine as not to be seen and that when broken initiate an electrical current which rings a bell in a guard house, warning of an enemy escape or in the case of a blockhouse crossing of the line. The inmates in Ceylon at least believed that this was used during the numerous crossings of De Wet, and had heard of this from their comrades.

Signal Rockets

Signal rockets were also used, initiated either by a remote striking device for rockets on the fence lines, or by lighting them from the blockhouse. Many were fitted just outside the gable-ended version, where the door could be opened and the rocket fired. These were to signal only, providing little or no illumination. They gave an alert to every blockhouse in the area, and many on hearing these alarms fired anyway, often creating a long-distance ripple of fire across the veld. The signal was also used to summon reinforcement from an armoured train where the blockhouse lines were in support of the railway line defences.

Ad Hoc Alarms

Many of the bully beef and biscuit tins once discarded were converted into rattles for the fences, designed to give a sound when the fence was touched. Some of these tin items can still be picked up off the veld, and came in many shapes and sizes, from complete tins to the end or pieces of dismembered tins, all fastened to the fence with more wire. On a windy night these must have rattled and flapped in the wind making a merry noise! Other alarms were also constructed out of simple rocks and trip wires designed to hit corrugated iron sheets, used particularly if there was any dead-ground along the wire.[205] In addition, alarm wires were often led to the metal roof of the blockhouse, which acted as a large sounding board to any tampering or cutting of the fence. Ingenious as these alarm guns and wires became, the Boers were equally cunning in overcoming them and rendering them largely ineffective.[206]

COMMUNICATION SYSTEMS

Visual Signalling

The high features of the country, and the almost perpetual sunshine during the greater part of the year, coupled with the clearness of the atmosphere and the great lines of sight, made South Africa an ideal visual signalling country. The main types used were the portable heliograph, using a mirror to send light signals as messages, similar to the larger lamps also used, and in addition to the more traditional semaphore flags shown here. The typical range for large signal flags was less than 5 miles (8 kilometres) and the rate at which a message could be sent in semaphore was only up to six words a minute for a highly skilled operator. This could be increased if silk flags were used (as they were lighter and with less wind resistance, they could be wielded much quicker!) and by converting messages to Morse Code.[207]

A signalling station supporting Colonel Burney's column in the hunt for Theron's commando.[208]

The heliograph – derived from the Greek *helios* (sun) and *graphos* (writing) – was a mobile signalling device using the sun and mirrors mounted on a tripod to signal in Morse Code to a distant site within eyesight and line-of-sight. Messages of up to 18 words per minute could be transmitted by an experienced heliograph operator. Reflectors came in different sizes; the larger the reflector the longer the line-of-sight distance that could be achieved. For example 5-inch (12.7 centimetres) had a distance of 50 miles (80 kilometres) while a 12-inch (30 centimetres) reflector could attain a range of about 80 miles (129 kilometres), subject to climatic conditions, of course. In fact, in other theatres at this time, such as India, distances of 100 miles (160 kilometres) had been achieved.

Although simple and easy to use – in fact a basic mirror might also be used to the same effect – the main disadvantage was that a light source was required, so that no use at night-time was possible and the messages could be intercepted along the line-of-sight unless encrypted. In addition, it could be difficult to use the heliograph when the sun was behind you, but there were drills and techniques to use a duplex mirror to overcome this. Similarly, there was also the wind to consider, but a sandbag could be suspended from the tripod to weigh it down by means of a handy hook, all covered in the *Army Signalling Manual* of the period.[209]

Each battalion and column had its own operators and they were typically used along the blockhouse railway lines as back-up to the telegraph system. This was perhaps the first use of an integrated communications system whereby a roving column might send messages into a blockhouse by heliograph, where operators could then convert and forward signal messages via telegraph. There are also accounts of bored blockhouse guards using this simple type of communication to send chess moves to other blockhouses in order to play a chess match! Such was the degree of their ingenuity once bored.

Rudyard Kipling also mentioned the use of heliographs in his poem 'Chant-Pagan':

> *Me that 'ave watch 'arf a world*
> *'Eave up all shiny with dew,*
> *Kopje on kop to the sun,*
> *An' as soon as the mist let 'em through*
> *Our 'elios winkin' like fun –*
> *Three sides of a ninety mile square,*
> *Over valleys as big as a shire –*
> *'Are ye there? Are ye there? Are ye there?'*
> *An' then the blind drum of our fire...*

A Boer War heliograph set.[210]

Equipment	Marching distance (range)	Speed (Words per min)	Comment
Electric telegraph	Limited only by length of line	300	24 hours service
Telephone	Limited only by length of line	200	24 hours service
Helio 12" reflector	83.5 miles (134 km)	8-16	Sunlight
Helio 10" reflector	83.5 miles (134 km)	8-16	Sunlight
Helio 9" reflector	83.5 miles (134 km)	8-16	Sunlight
Helio 5" reflector	52.5 miles (84.5 km)	8-16	Sunlight
Helio (saddle) 3" reflector	37 miles (59.5 km)	8-16	Sunlight
Helio (limelight) 5" reflector	31 miles (50 km)	8-16	
Helio (moonlight) 5" reflector	12 miles (19.3 km)	8-16	Strong moonlight
Lamp, trench pattern shutter	Short and variable	4	
Lamp, electric, daylight signalling	Not stated, estimated 30 miles (48.2 km)	8-16	
Flags, Large	Difficult over 5 miles (8 km)	2	Faster with silk flags and Morse code, slower using semaphore

Anglo-Boer War communication equipment with range and message speed performances.[211]

Physical Dispatches

The use of runners to send physical and verbal orders across the battlefield was still important during this war, especially at the lower levels within units. There were also dispatch riders both on horseback and in bicycle units to pass detailed messages and orders across the wider expanse of the African veld.

Telegraph and Telephone

Visual signalling along the blockhouse lines was to a large extent superseded by telephone and telegraph towards the end of the war. Keeping the blockhouses and the lines in contact with their own regimental headquarters was vital, not least for reinforcement when attacks occurred. The armoured trains that were on call for reinforcement also acted as mobile telegraph offices, and were connected into the telegraph system when static.

As the blockhouse system grew from the defence of the key points and the lines of communication rail links, so did the communications that linked them together. Initially when the large masonry blockhouses defending key bridges and towns needed to be able to call for reinforcements should there be an attack, these would usually come from one of the armoured trains. Later in the war, when the small Rice Pattern blockhouses were deployed cross-country, they needed communications in order to co-ordinate the drives, and for local fighting when the Boers tried to breach the lines.

Line was laid along all the rail tracks, either underground or on telegraph poles referred to as 'air lines'. This could provide either signalled telegraph or telephone communications, but was vulnerable to being cut and to eavesdropping by roving Boer commandos. Typically each fourth blockhouse would have a telephone and be linked in to its subordinate headquarters in a hierarchical configuration. These would then be connected via their regimental and divisional networks to major headquarters and signal message centres, such as De Aar, the railhead for the dispersal of the British forces, and Pretoria, the British headquarters. Approximately 370 telephones were installed in blockhouses during the war.[212]

Due to the lack of wireless radio communications at this time, the extensive use of both telegraph and telephone was vital in order to co-ordinate the drives, armoured trains and blockhouses lines. There was therefore an extensive network of communications laid out across the theatre of operations as shown in this table from the Army Telegraph Report for 1900-1902:

Cable Type	Miles	Use
Heavy iron Wire	830	Railways lines
Light iron & copper Wire	600	Railways lines
Telephone and telegraph communication lines	800	Cross-country 520 miles
Temporary lines laid in the Cape Colony	700	
Total	2,930	

Army telegraph report 1900–1902 showing amount of cable used.

Fortifying the railway lines with blockhouses required the laying of a great number of telephones for inter-communication. As a rule, a wire was run out each side of a blockhouse, acting as a control (at which there would be a telegraph office) and five telephones placed on either, to respective blockhouses. It was found that five telephones on a circuit represented the maximum for efficient working.

Local interception of these lines led to the use of deception messages sent in clear text to mislead the enemy and then to the use of cipher to encrypt messages and provide security to the passage of orders. In addition to these static forms of communication for the blockhouse lines each unit had its normal flag, lamp and heliograph communications, which although primarily deployed on mobile operations for local communication, could also be used as a backup in the defensive role and for communications to the roving columns should this be required.

Wireless Radio in the Boer War[213]

The war heralded onto the battlefield for the first time the use of wireless radio to transmit Morse-coded telegraph messages, to various degrees of success, by the British Army and the Royal Navy. It was the interests of the Navy and the outcome of their manoeuvres in the summer of 1899 that were instrumental in the War Office's decision to procure Marconi's wireless apparatus and send it to South Africa. A small contingent of six Marconi engineers and approximately 12 RE Sappers equipped with five 'portable' wireless stations under the command of Captain JNC Kennedy RE arrived in Cape Town in December 1899.

The original deployment concept, born from naval trials in Britain, was to control the disembarkation of troops using ship-to-shore links, and to coordinate the process of landing masses of men, animals and materiel. However Captain Kennedy, perhaps keen to demonstrate the new technology that had been previously snubbed by the Army, offered to give the generals and the staff a demonstration in the grounds of the Castle of Good Hope in Cape Town. Although only demonstrated over a few hundred metres, the impressed staff officers decided to divert the five wireless sets from naval operations to the front instead. There is a plaque commemorating the first use of wireless telegraphy in the Castle of Good Hope, from where the radio operations were run during the early stages of the war.

Transiting via De Aar, the wireless contingents deployed three outstations to Orange River, Belmont and Modder River, with the main station being sent to Lord Methuen's headquarters at Enslin. One set remained at De Aar. Unfortunately the new technology depended on tall masts on which to erect the antennas and these had not been adequately provided. So high was the requirement that kites and balloons could be used, depending on the weather, although the ad hoc preference was to lash bamboo poles together to reach a reasonable height. The deployment however was not successful and it took over a month to send any intelligible messages and only over a distance of 50 miles (80 kilometres) with an intermediate relay. After a further six weeks, and being able to offer no credible reasons for failure, the sets were dismantled and sent to the Royal Navy at Simon's Town, where Adjutant-General Sir Evelyn Wood believed they might have some use for them.

The wireless equipment in theatre was not unique at the time, however: President Kruger had been busy acquiring wireless radio as part of the Boer militarisation programme. After competitive tenders from Marconi, Siemens and Halske, six sets of *vonkeltelegrafinstrumenten* (Spark telegraph instrument) equipment were procured. By the time they arrived in Cape Town the war was already in progress and the sets were impounded by Customs and never made it to the Boers. Captain Kennedy, in true signals style, cannibalised the sets for spares for his own equipment, but discarded the steel masts required for the antennas – not that any of it seemed to make a difference getting Marconi's equipment to work.

The Royal Navy however had much greater success deploying the sets onto ships to support the blockade of the Delagoa Bay area preventing supplies getting to the Boers via Lourenço Marques. Ship-to-ship communications worked well over great distances, aiding the command and control of ships over the horizon for the first time in a theatre of war. The success in this operation led the navy to equip 42 ships and eight shore stations round Britain with wireless by the end of 1900. The era of modern battle-space communication had begun. Such was the magnitude of this first deployment of radio wireless technology that the South African Institute of Electrical and Electronic Engineers made the following historical milestone declaration: 'The first use of wireless telegraphy in the field occurred during the Anglo-Boer War (1899–1902). The British Army experimented with Marconi's system and the British Navy successfully used it for communication among naval vessels in Delagoa Bay, prompting further development of Marconi's wireless telegraph system for practical uses.'[214]

Armoured Train Telegraph

The armoured trains used for reinforcement along the railway blockhouse lines could only communicate when stationary and had a small contingent of three signallers who could connect to the telegraph wires. The static nature of these land-laid line connections proved to be a significant constraint to command and control and one that would continue well into the next century. During the First World War some 15 years later, line was continually being laid, cut and re-laid for the duration of the war. Armoured trains were officially recognised as moving telegraph offices, and equipped with field sounders, vibrators, phonophores and telephones; and when trains stopped away from a regular office, which they did nearly every night, they were never out of communication with the neighbouring stations and blockhouses.

Wireless Radio

It was during the Boer War that the first introduction of wireless technology was pioneered on the battlefield. Marconi suggested the deployment of his new wireless set to demonstrate the technology and under the direction of Captain JNC Kennedy used sets to provide communications for some naval deployments. The notion of technology demonstrators is still evident in today's armed forces, especially with the increasing complexity of communications needs and commensurate threat. It is of note that where the technology was most needed then, it was perhaps least provided; at the lowest level, amid the 'fog of war', in the middle of the night during a Boer attack, it might have produced the best form of control. Continuing the theme forward in time, it was not till the late 20th century that the lowest combat sections of the British Army received fully secure wireless communications in the form of the Bowman Radio System.

Royal Engineers/Marconi Company wireless section at De Aar encampment, South Africa, 1899.[215] Courtesy of Royal Signal Museum, Blandford.

SUPPLY SYSTEM

As the blockhouse system grew to its final contingent of over 50,000 British and Colonial troops and 25,000 Africans in residence, so the logistics problem also steadily grew. Soldiers had to be fed, watered and rationed in a demanding climate and terrain, which placed a strain on the depots of supply and regimental echelons to provide transport on a continuous basis. As blockhouses were deployed in accordance with a regimental hierarchy, the larger company blockhouses provided a base from which to resupply the units' subordinate blockhouses. This distributed system meant that particularly along the railways, trains did not have to stop every half mile to resupply individual blockhouses, but could be distributed at stations or unit headquarters. This supply chain system was issued as an order by the theatre Adjutant-General, Major-General WF Kelly, in December 1901,[216] stating: 'Posts and Blockhouses between stations are not to depend on the train service for their daily rations and water.'

Blockhouses were usually supplied for a fortnight, with a reserve supply both of ammunition and food to be used only in cases of emergency.[217] In addition, and particularly when establishing the blockhouse lines during construction, any local farms or dwellings were ripe for foraging or pilfering. Everything was in scarce supply and the rounding up of cattle and horses from farms would no doubt have meant the liberation of goats, pigs and chickens for a fine supplement to the issued rations. Fresh eggs were a bonus and would have made the blockhouse particularly interesting to visitors, such as commanding officers and the padre.

Daily delivery from the regimental water cart.
Courtesy of National Army Museum, London.[218]

Water was a particularly important requirement and the Royal Engineers were tasked to drill boreholes at a staggering rate to keep up with demand. Water had then to be stored in barrels and transported on carts to the various locations. Clean water was a battle-winning requirement, as the spread of water-borne diseases had caused significant casualties earlier in the war. Each blockhouse was specifically provided with one or two water tanks for storage and encouraged to collect rain water as a supplement to the water delivered by cart.[219] These water tanks were often dug in for protection from fire and from the sun's baking rays or placed with the masonry blockhouse on the ground floor.

Africans Supporting the War Effort

Both sides had at least an initial gentleman's agreement that this was their 'White-man's War' and that the black population should not be brought into the war as combatants. Inevitably as the war became protracted, each side not so reluctantly used the local people in their war effort. Initially the British used locals as a ready source of labour in supporting unarmed roles; as farriers, blacksmiths and wheelwrights; to build the railways; to drive wagon trains and to construct the blockhouse lines. It is estimated that 100,000 locals were used throughout the war in these roles. The Boers had more intimate support while 'on commando' and their farm workers became their *agterryers*, guarding, cooking and reloading firearms.[220] These men were armed for self-protection only, as the British would also do with their scouts used for reconnaissance roles later in the war as it became more mobile. Ultimately they were drawn into the armed struggle on both sides and caught between two warring powers.

The Boers were fearful of the Africans, having previously fought several wars with them, and wanted the British to ensure they remained unarmed and neutral. In some cases, for instance, the Swazi wanted to settle old scores with the Boers, who had confiscated their land before the war,[221] while Zulus also had old scores to settle from previous conflicts. In addition, during the war roving commandos raided African villages for provisions. Finally, the Africans were abjectly poor: those who worked for the transport department of the British Army as drivers or sentries, for example, were paid as much as £3 per month, as opposed to the Boers, who were commandeering the black workforce and paying nothing for their labour.[222]

A contingent of 'native blockhouse guards'.[223]
Courtesy of National Army Museum, London.

Kitchener finally decided to use armed Africans in the blockhouse lines and over 25,000 served in this capacity, some in blockhouses themselves, and others as night-guards or for observation posts. It is estimated that a further 15,000 were armed and served in the mobile columns as scouts and auxiliaries. Although Kitchener was reluctant to admit the extent to which he had armed Africans to a wary Cabinet, he was not deterred by racial prejudice from making the best use of 'native allies' in his unrelenting drive for victory. In the Sudan, he had also 'gone native', learning Arabic, and since then had collected Arabic art; he would use any means to achieve his aim.

'Native African scouts employed by the British Army'.[224]
Courtesy of the National Army Museum, London.

In addition the British sought alliances with local tribes who had previously fought the Boers, such as the Pedi and Barolong, encouraging them to raid Boer laagers and deny wagon routes to their supply echelons. Units of Pedi, Swazi, Shangaan and Tsonga tribesmen were also raised, and served with units such as Steinaecker's Horse, to patrol and prevent Boer arms reaching the Highveld. Often the spoils of these raids would be divided up between the British and the tribes concerned.

On the other hand, Africans were also victims caught up in the scorched earth policy and were swept off the veld with the remaining Boer women and children of both loyal burgher families and those on commando. They were interned as servants housed with Boer families in the white camps, but also had their own camps established on a similar scale, where some would also serve as medical orderlies and stretcher bearers. These camps had much worse conditions and were largely unrecognised in terms of the suffering endured; none of the Women's Committee who visited from England even set foot inside one of the black camps. By the end of the war, thousands of black inmates had died from typhoid, diarrhoea and dysentery because of the appalling conditions under which they had lived. The camps remained under British military control throughout the war and lacked the elaborate medical and educational structures of the white camps. By the end of the war at least 115,000 had been interned in black camps,[225] with another 4,000 in white camps. The official number of recorded deaths was 14,154, but this was grossly underestimated; G Benneyworth estimates it as at least 20,000, after examining actual graveyards.[226]

A photograph from the National Army Museum, London archives shows a 'medical orderly' in one of the concentration camps. What he is collecting or delivering with his cart is not specified, but he looks to be officially dressed and under employment. The photograph was taken by Miss MS Barwell who was both a prolific photographer in the Anglo-Boer War, having taken over 100 images, and also mentioned in despatches by Field Marshal French in 1914–16, still nursing the sick years later.

An African medical orderly at a British camp, c1901.[227]
Courtesy of the National Army Museum, London.

In the Cape Colony, the situation was summarised in a letter to the Secretary of State for War as follows: 'In the cape colony; Cape Boys and Bastards are separated by act of parliament from natives, of the latter we have some in blockhouses though we have some as watchmen between blockhouses; of the former Cape Boys, I believe French has actually allowed some of them to occupy blockhouses out west. Of the Bastards [sic] there has always been a corps at Upington on the German frontier to guard roads and well holes. They may be said to be police.'[228]

Black, Coloured and Indian participants were barely recognised for their service and suffering during the conflict and after the war. The British only honoured White combatants until changes to the new democracy in 1994 brought to recognition the suffering that occurred in the Black 'internment camps', and steps were taken to provide graves and memorials to them.

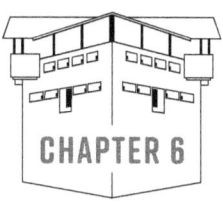

CHAPTER 6

MOBILE BLOCKHOUSES

*Strong could be dealt with—but crazy? Oh no.
You never knew what the crazy were going to do.*

ANDREA SPEED, *SHIFT*

Although blockhouses were traditionally static fortifications, they were also required to protect the mobile logistics columns and baggage trains supplying them. So came the 'mobile blockhouse' (or 'movable' or 'travelling' blockhouse).

Some have heralded the mobile blockhouse as the forerunner to the tank, as its nature was somewhat tank-like, having iron-clad protection, firepower from the loop-holes and mobility from the drawing animals; the three classic capability trade-offs of the modern-day battle stalwart. However, it is the nature of the mobility gained from animals, either horses or oxen, which for the author places this in history, not as the first tank, but as an armoured wagon. Clearly the first thing any attacker would do to disable the wagon would be to shoot the drawing animals, and it is the lack of an internal or integrated motorised component which denies it the honour of being the first tank.

In Willem Steenkamp's book, *The Black Beret*,[229] he writes about the 37mm pom-poms mounted on armoured ox-wagons as early as the 1890s and also later during the 1899–1902 war. Prior to even considering static or mobile ox-driven blockhouses, however, the British had identified the need very early in the war to armour their steam traction engines and developed a 'road-train' system that most likely was the inspiration for the mobile ox-driven versions. The need in this case was to protect the road supply routes from attack by the commandos, and the Royal Engineers companies also operated these armoured road trains.

ARMOURED STEAM ROAD TRAINS

Steam was still a modern mode of transport at the turn of the century and Britain was the largest producer of steam traction engines. The first steam tractors were built in the 1850s and used in the American Civil War to good effect, after which their use proliferated. Many armies experimented with their use, which was predominantly for road train supply on well-prepared roads. They could move military materiel at three times the speed of animal-drawn wagons, a rather quick 10 mph (16 kmh).[230]

The British Empire was a prolific user of steam, especially in remote areas in Australia and India. In South Africa its use on the outbreak of war was immediate on both sides. Available steam tractors were pressed into use, and the War Office bought more in 1899 to send to the Anglo-Boer conflict. By the end of the war steam road transport in South Africa comprised 46 traction engines and 730 men.[231] Operating the steam traction engines was the responsibility of the Royal Corps of Engineers. One such unit, 45th (Fortress) Company RE, had been raised in September of 1899 as a Fortress Company but was given the special mission of serving as a Steam Road Transport Company. The company was equipped with steam traction engines manufactured by three firms, Fowler, Burrell, and McLaren.

John Fowler and Company, of Leeds, England was requested to build a fleet of seven armoured steam traction engines, and twenty support carriages in 1900.[232] Called the Fowler B5 Armoured Road Tractor, they had a 20 horsepower steam engine surrounded by an armoured shell to protect it and the crew. The standard B5 engine weighed 8.5 tons, but with the 5/16 inch (8 millimetres) armour fitted (supplied by Charles Cammell & Co Ltd) the tractor was nearly 23 tons (21 tonnes). The engine's normal speed was 10 mph on level, improved roads and at this time it had no rough country or off-road capability and was limited to well-prepared hard roads. However with the 4.5 tons (4.1 tonnes) of armour fitted this reduced to a maximum speed of 6 mph (10 kmh).[233] The range was only limited by how much coal and water it could carry to service its boiler or could be delivered to it, and was usually around 10-17 miles (16-27 kilometres).[234]

The prototype shown in the photograph is the first of a series of armoured engines that underwent tests during May 1900 in Leeds for the War Office prior to being shipped to South Africa. The trials however were not well regarded and the engine's performance on soft or inclined ground was poor, as reported by the invited trade press. It was an indication of the limited off-road capability while fully laden, although during the trial the engine was able to decouple its carriages and use its powerful winch to haul itself and the carriages up the slope, albeit incurring a delay.[235]

A class B5 'Super Lion' model, special armoured road locomotive No 8894, built in early 1900, with armoured wagons and two guns. The train comprised a haulage engine weighing 20 tons (18 tonnes) and three armoured wagons to carry ammunition for two 6 inch howitzers weighing 60 tons (54 tonnes).

Like the railway trains, the road transport system and the convoys of men and materiel moving by steam tractors were themselves attacked and sniped at, and there soon arose a need for convoys to increase their armoured protection. The general-purpose traction engines were used for pulling trains of supplies, alongside oxen or mule transport, or towing guns to different positions. For protection against attacks on supply columns by Boer raiding parties some armoured traction engines were ordered, and the first of these vehicles arrived in South Africa in July 1900, followed by a second train two weeks later. The engine and carriages were fitted with loopholes in the steel-plated armour for the crew's rifles, similar to the mobile blockhouses built later in the war.

The first two road trains (Nos 8894 & 8895) were delivered to Bloemfontein in May and June, but the armour was removed from both engines and carriages and used to make armoured railway trains. By the end of 1901 the General Officer Commanding (GOC) Kimberley District asked for further trucks to be fitted with armour so that the troops needed for road-convoy escort duties could be reduced. The War Office was requested to supply two armoured trucks; however, remembering that the first two sent had been stripped of their armour to make armoured railway trains, it is not surprising that they did not meet this request. By the end of 1901 there were some 45 steam transport trains in service in theatre.[236]

The remaining five traction engine trains arrived later in July, August, October (two) and December, and there is no evidence to suggest that these were used in any other theatre, making them unique to South Africa. There is also no evidence that they saw any action at all as armoured engines or carriages. Rather they were stripped for their much-needed armour plates to equip the armoured trains. They were then used in the service of the Imperial Military Railways as ordinary engines

It is not known how they were disposed of after the war ended, as none seem to have survived. At least one standard Fowler B5 (No 8726) was recovered in a South African scrapyard and restored. It used to reside in the North of England in the Vintage Vehicle Museum in Shildon, County Durham, but this closed to the public in 2012.[237] Others may have remained in South Africa and been put to work on the mines, possibly in the Springs area, and one more at least made it back to the UK where Fowler 8894 was sold as a wreck at auction in 2012.[238]

Steam engines were still utilised in the First World War, and some examples up to the Second World War, but then quickly disappeared as the internal combustion engine took over. Having arrived so early in the campaign these steam engines would have most likely provided the inspiration for the mobile blockhouses.

Fowler 5B engine and carriage detail. Courtesy of the Museum of English Rural Life, University of Reading, John Fowler & Co (Leeds) Ltd Collection.[239]

1. Rear view of the class B5 Lion special armoured compound road locomotive No 8894, showing the rear door open. The entire traction engine was enclosed in a steel box, with only the smoke stack penetrating the armoured skin. Cammell Laird & Co rated the main armour as providing protection against both the British Lee-Metford and the German Mauser at a range of just 20 yards (18 metres). Given that Boer ambushes were usually from positions at 200 yards (180 metres) such protection could be seen as excessive, and better reduced to ¼ inch (6 millimetres) steel plate.

2. An armoured traction wagon or carriage, one of a batch numbered 4945–68, built in 1900 for Fowler by Charles Cammell & Co. This view is the rear nearside with the loading doors closed. The view shows the slightly angled loopholed sides in a vertical position, which could then act as firing positions for the troops inside. The sides could be lowered to form a lower enclosed space in which to transport vulnerable stores such as ammunition. If unhitched a wagon could be used as a static blockhouse and this notion may have contributed to the later ox-wagon drawn blockhouses.

3. An armoured traction wagon. This view is the rear with the doors open showing a single 6 inch howitzer loaded inside, as opposed to towed behind it. There is also a steel lattice roof over the wagon which could have had a canvas fitted during transit to protect the load or troops.

MOBILE OX-DRIVEN BLOCKHOUSES

Several different designs for ox-driven blockhouses were initiated, with the first reportedly manufactured in 1901. Although it is widely accepted that the ideas and designs for the mobile blockhouses came solely from Engineer sources, there were also others who were beginning to think of mobile armoured positions. One such proponent was Colonel HS Rawlinson, a favourite and protégé of both Roberts and Kitchener, who considered him one of his best column commanders. Between April and December 1901 his columns of nearly 1,500 men had marched 3,400 miles (5,470 kilometres) and accounted for a mere 538 Boers killed, wounded or captured and sent to POW camps. This was considered successful, but at a significant price.

Rawlinson seemed to understand the need to further compartmentalise the vast tracts of land, beyond that already achieved. By mid-1901 the railways were already developing into formidable barriers but there was still a need to capture the remaining Boer commandos roaming the veld. As a member of the 'Kitchener Ring' Rawlinson felt able to write to him directly from the General Staff perspective with this suggestion of a movable blockhouse line based on fixed points and mobile ox-wagon blockhouses (see opposite).[240]

In the latter stages of 1901, Royal Engineer units such as the 38th (Field) Company RE based in Bloemfontein had already started to produce some mobile blockhouses. Also the 47th (Fortress) Company RE had by this time already deployed two movable blockhouses for operational use on Colonel Sir Robert Colleton's column in the Johannesburg area. Although the design may not necessarily be unique, its usage and operational deployment as part of a semi-fixed and movable line was a new concept for the period. Rawlinson however alludes to the fact that wagons are already available at some of the depots, revealing that he may have taken an existing design and incorporated it into a tactic for the veld.

COMMAND-IN-CHIEF'S OFFICE
SOUTH AFRICA
RAWLINSON   1901

My Dear General

This fierce country we are now in I fear too big to do very much good in until you have some more blockhouse lines out. I know you do not want to erect permanent lines but could not something on the enclosed lines to move with the Seaforths, Camerons and Argyles? If Delaney went off to the Marico the Highland Brigade with its armoured wagons might easily come up there or anywhere else that you might require them. I think the wagons are probably available at some of the depots and any cattle would go to fit in them.

    This is only an idea, which the SO Blockhouses would be able to improve upon. If it could be done it would be extremely useful in this base open veld and with the Highland Brigade holding 150 miles of Blockhouses we should be able to divide the country very satisfactorily.

Johnny Hamilton* arrives this afternoon,

Yours ever,

Rawly

General Ian Hamilton, Kitchener's Chief-of-Staff, known as Johnny, arrived to meet Kitchener in Pretoria early December 1901.

Pages 1-2

A MOVABLE BLOCKHOUSE LINE

The unit would be an armoured wagon as below with sufficient plates (loopholed) for a second blockhouse about 800 yards distant. This wagon would carry one mile of wire with standards complete a 20-gallon water tank and the kits, tools and ammunition for 16 men including one NCO.

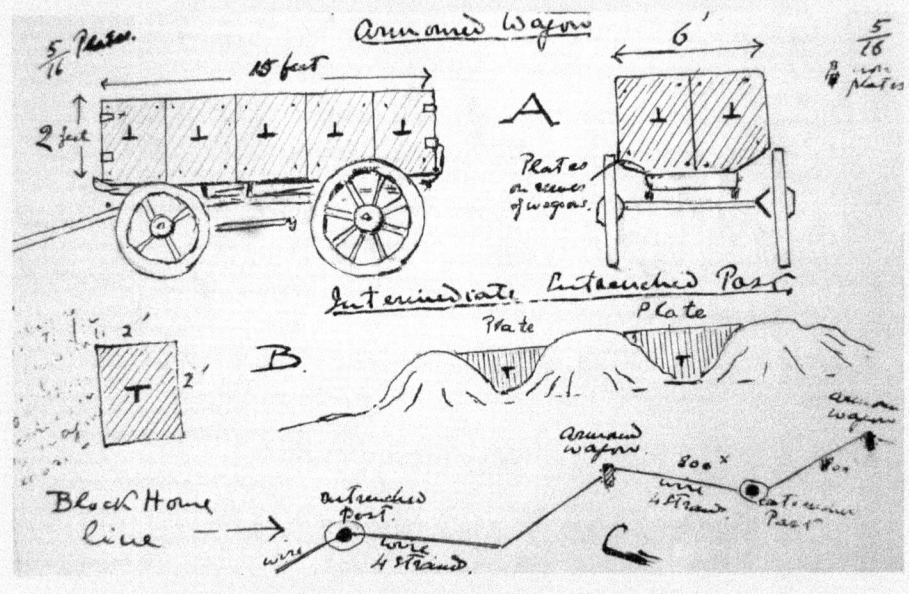

Armoured wagon.

Side pencil notes read: Armoured plate for intermediate post, 10 of these carried in the wagon – loose.

Page 3

The weights would work out something as follows
for each wagon.

1 mile of wire 4 strands with up rights complete	3,000 lbs.
Armoured plates for one wagon as shown above as sides of wagon and at both ends of same giving lying down protected cover and men could fire over the top at night kneeling	2,000 lbs.
10 loopholed plates for the entrenched work between each wagon	1,000 lbs.

Water tank (20 gallons) tools, kits and ammunition
for 16 men   1,000 lbs.

Total   7,000 lbs.

For each 16 men one of the above wagons would be necessary any and they would hold one mile of blockhouse line wound up and safe.

Therefore for a Battalion 800 strong 50 wagons would be required and the battalion hold 50 miles of blockhouse line.

If oxen could be spared at the rate of 10 per wagon the line could at any time take down or erect their own mile of wire and complete their own entrenched post simultaneously. With the assistance of a few mounted troops they might even do a drive ! on their own account at the rate of more than 8 or 9 miles a day.

Pages 4-5

Letter from Sir HS Rawlinson to Lord Kitchener suggesting blockhouse lines and mobile blockhouses on wagon, dated 1901.

The mobile design finally produced was based on a four-wheeled ox-wagon similar to that of an armoured train or armoured traction engine. Initially they were manufactured in Middelburg and Pretoria with the design claimed by Elliot Wood.[241] They comprised four 1/2 inch (12.5 millimetres) steel plate walls, each with T-shaped loopholes. The roof comprised tent material, allowing the wagon to be used covered or uncovered depending on the climate. These hooped roofs also allowed the sides to be dropped so that they looked like peaceful supply wagons as subtle camouflage.

The wagons were heavy, approximately 1.36 tons (1,800 kilograms). The Pretoria garrison, however, constructed a lighter one made from Pretoria Fort window shutters using 3/16 inch (4.8 millimetres) hardened steel, which greatly reduced the weight to 1,500 pounds (680 kilograms). This weight reduction proved to be popular and the hardened plate was ordered from England to construct light mobile blockhouses for mule transport with the convoys marching off the blockhouse lines. The records do not show if this ever arrived in time to be used in mobile blockhouse construction.[242]

Ox-wagon blockhouses[243] constructed by 47th (Fortress) Company RE. Note the two different types, two foreground of steel plate, and the one in the rear made of corrugated-iron. Reproduced by permission of Durham University Library and Collections.

Typical tasks for these mobile variants and others like them included:
- Convoy protection duty.
- Reinforcing the existing blockhouse lines, to supplement armoured train operations.
- Providing a stop-gap where the Boer forces had already made an incursion, until the permanent line was repaired.
- Static defence for blockhouse construction parties, usually at the construction head.
- Mobile supply protection on good traversable roads.

Reportedly, one regimental colonel had his ox-wagon fitted with a one-roomed corrugated iron house for his living quarters – after all, this was supposed to be a 'gentleman's war'![244]

The new year of 1902 found the British connecting Ventersdorp with Tafelkop by a blockhouse line. It was in this area at the beginning of February 1902 that Colonel James Francis Hickie and Colonel Robert Kekewich combined their forces and on 4 February made an attempt to surprise General De la Rey in the area south of Klerksdorp. Operating within a triangle of blockhouses and the railway line, which ran from Klerksdorp to Ventersdorp and from Ventersdorp to Potchefstroom, they would have supplemented their operations with these mobile blockhouses. In another encounter, hunting De Wet where mobile blockhouses were not available, inventive officers improvised by setting a chain of pickets, 'each post having a wagon banked up with forage and mealie sacks to serve as a movable blockhouse'.[245]

There are limited references to these mobile adaptations, but sources from the National Army Museum in London show the construction of two very different types. The first one easily shows its roots in the mass-produced Rice Pattern blockhouses, much smaller in diameter in order to fit onto the short axle width of a horse-drawn supply wagon, which was about 2 metres. Developed later in the war around March 1902,[246] most likely the standard parts of the Rice Pattern were used but their bend arc increased by continued use of the rolling machine, so as to achieve the smaller diameter required. This could be done manually using a hand-operated roller, like the one on display at the War Museum in Bloemfontein.

*A mobile Rice-type blockhouse under construction.
Courtesy of the National Army Museum, London.*[247]

The model above was designed by 47th (Field) Company RE and based on an ox-wagon chassis. It had an outer corrugated iron wall, similar to the Rice patterns, but used a timber frame The intervening gap was filled with either shingle or sand, with a total weight of between 4,500 pounds and 7,000 pounds (2.0–3.2 tonnes), and was designed to be pulled by a team of at least four oxen.[248]

In the following photograph, the frame has some sort of sitting arrangement or steps to allow a few men – possibly a maximum of four – to climb in. It is unclear how thick the walls were, but they must have been only a few centimetres, due to space and weight limitations of the wagon, questioning the effectiveness of such a vehicle. This type of mobile blockhouse was used to protect the convoys and also could be used to augment the blockhouse lines where there might be an increased threat of the Boer commandos crossing over.

Two different types of mobile rectangular blockhouse mounted on ox-wagons. Courtesy of the National Army Museum.[249]

The second mobile type shown here is a much larger variant, covering the entire flat bed of the ox-wagon. It has 12–14 loopholes, catering for a much larger contingent of troops, maybe up to 14 in total. Even made out of single-thickness iron or steel plates for protection, these would have been extremely heavy, although some weight is saved by having a canvas pitched roof, which would have allowed the incumbents to stand at the loop-holes when required. In both these examples it would have been an uncomfortable ride. Passenger fatigue in all-terrain armoured vehicles was one of the issues for the first tanks in the First World War and continues to be an extremely important part of vehicle design today (now termed Human Factors and more commonly known as ergonomics).

Another use for ox-wagons, although not armoured, was for the mounting of searchlights, as remarked in one of the Royal Engineers War Diaries for Natal: 'A searchlight was organised from apparatus taken out of one of the tugs in Durban Harbours. It was mounted on an ox-wagon, and proved so satisfactory that it was sent to the front.'[250] Such was the adaptability of these wagons.

The opposing Boer forces also adapted ox-wagons over the same period, of a similar design to the full-length blockhouse previously shown.[251] Reinforced by timber and steel plate, it was most likely an isolated experiment or conversion, as this type of defence would have been contrary to the light mobility of the commandos. Although they did have a ready supply of ox-wagons the armouring of these would have severely slowed them down, making them more vulnerable to capture.

Key disadvantages were that they were heavy and slow to manoeuvre, subject to problems of hilly or wet terrain, and their oxen draughts required fodder and water in order to retain mobility. Elliot Wood in his account after the war states that they were mainly towed by motor transport (steam engines) due to the weight. This would have also meant they were restricted to deployment in areas close to suitable roads.

For both forces these early experiments heralded the first armoured fighting vehicles, albeit animal-drawn; the next step was full mortised mobility and the dawning of 20th century mobile warfare. The key aspects of armoured manoeuvre warfare were already being played out in South Africa with these mobile blockhouses.

A depiction of a mobile blockhouse crossing a pass, (note this example is made of corrugated-iron panels).[252] It may be a stylised drawing as no photographic evidence exists of this design, although protection of convoys was a task.

The only other type of armoured mobile vehicles during this war, other than the armoured trains, was the steam engine or traction engine, used for logistics supply. These were commercial vehicles, which used steel plate to protect the driver from small arms fire, and which ran mainly between Johannesburg and Pretoria ferrying supplies.

At some point there was even an experiment made with camouflage. Although there is photographic evidence, no reference is made to this in any of the engineer diaries of the period or the post-war reports. Maybe it was a novelty, or bored soldiers having some entertainment. Either way, it seems to be a unique 'once-off' for the war and there are no other examples of camouflage for vehicles. The design is oddly reminiscent, for the author, of a novel urban camouflage pattern used in Berlin in the 1980s, where similar angular squares were used in urban colours for tank camouflage. Maybe this blockhouse pattern had similar aims.

An early experiment with checker-board camouflage.[253]
Courtesy of SANDF Document Archive.

Although there are no remaining examples of the British mobile blockhouse and little documentary evidence other than photographs, they are commemorated on the Royal Engineers Memorial Arch to the war built in their corps barracks in Chatham, Kent. One of the sculptured panels, which are said to be based upon photographs of the actual incidents and scenes of the time,[254] depicts "two moveable ox-wagon blockhouses". Of the two designs commemorated one is of the steel plate design and the other of a corrugated iron sandwich wall construction as used on Kekewich's Column, both shown with canvas covers for camouflage and protection from the elements.

In the move to mechanised armoured warfare in the early part of the 20th century the Americans also developed the Dayton Mobile Blockhouse in 1903.[255] This practically unknown American prototype Vehicle-Fort was designed and manufactured by the Engineer and Captain Edwin Dayton of the 22nd Regiment of New York. Its design corresponded to the need to create, quickly, strong points in New York's streets, against hypothetical severe public disorder.

FIRST WORLD WAR

The advent of the first tank was many years later on 15th September 1916 at the battle of Flers-Courcelette with the unique rhomboid shaped Mark I Tank; its German counterpart, the AV7,[256] appeared some two years later in the spring of 1918. The first use of tanks en masse occurred at the Battle of Cambrai in November 1917, when approximately 460 Mk IVs were used. The first tank-to-tank battle was between Mk IV tanks and German A7Vs in the Second Battle of Villers-Bretonneux, April 1918. No types of movable blockhouse were seen during this war, most likely due to its static nature and the heavy use of artillery, which made the use of such vehicles meaningless.

SECOND WORLD WAR

The concept of mobile but 'parkable' fortifications or defensive vehicles was also taken into the Second World War, where 'The Bison' was deployed by the Home Guard to defend airfields used by the Royal Air Force (RAF). They came to prominence during the period 1940–1941, when there was a strong fear of potential invasion by the German Army. There were a number of variants built on existing long-wheel-base lorry chassis, having a large rear compartment protected by the unique use of prefabricated concrete blocks.

The fear of a paratrooper landing at the RAF airfields prompted the deployment of tanks or armoured cars in this defensive role, but they were in short supply after the retreat at Dunkirk. Essentially The Bison provided an alternative that was relatively easy to build and could be easily driven around an airfield or used to block runways.

The Bison was the invention of Charles Bernard Mathews, who was a director of Concrete Limited, and approximately 250 were produced from commercially available and existing lorry chassis, often from old First World War vehicles such as fire engines. The protection took its lead from the concrete pillboxes constructed during the same war. There were three variant builds as follows:

- Type 1 was the lightest. It had a fully armoured cab and a small armoured fighting compartment roofed with canvas.
- Type 2 had an armoured cab roofed with canvas and a separate fully enclosed fighting compartment resembling a small pillbox on the back – communication between driver and crew must have been difficult.
- Type 3 was the heaviest and largest, with a contiguous cabin and fighting compartment completely enclosed in concrete armour.

The complete Type 1 Bison[257] concrete armoured lorry shown here (left) was reconstructed from extant parts and a period lorry chassis, on show at the Bovington Tank Museum, Dorset, UK. This example is a Thornycroft Tartar 3 ton, 6x4 of 1931 or later. The chassis here was a military forward-control Tartar, used as a 3-ton general service lorry.[258] (Courtesy The Tank Museum, Bovington, Dorset.)

The Type 3 Bison[259] shown here (below) is very reminiscent of the ox-wagon version used some 45 years previously. Oxen have now been replaced with a vehicle engine.

Bison Mobile Blockhouses, WWII.

Date	Built By	Qty	Type	Where / When Deployed
January 1901	23rd (Field) Company RE (Middelberg)	1	A steel-plated travelling blockhouse on a wagon was completed	Not stated (REC BW 8)
March 1901	23rd (Field) Company RE (Middelberg)	2+	Quantity not specified for Rice Circular Pattern and Steel-Plate Pattern (Plans only)	Used for 'slow work and where roads were not too bad'. (REC BW 8)
October 1901	47th (Fortress) Company RE (Johannesburg)	2	Two mobile blockhouse (4.8mm (3/16 inch) Steel Plate)	November 1901, they were deployed with Colonel Sir Robert Colleton's Column protecting blockhouse construction parties.
During 1901	38th (Field) Company RE (Bloemfontein)	2+	At the suggestion of the GOC "some movable blockhouses were made, by erecting a kind of Rice Blockhouse on ox-wagons"	Not stated (REC BW 11)
Dec 1901 – May 1902	CRE Kimberley District 32	7	"4 – with steel plates 1 – with CGI & shingle 2 – SRY wagons with steel plates"	(REC BW 3 CRE Report)
January 1902	Tany Ho. Contractors 47th (Fortress) Company RE (Johannesburg)	2	Two mobile blockhouse (16mm (5/8 inch)) steel-plated), as no ½ inch plate was available	February 1902, Sent to Klerksdorp for use on the line to Lichtenberg
February 1902 (1st Quarter)	47th (Fortress) Company RE (Johannesburg)	1	One movable blockhouse built to Rice "Principles" weighing 4,500 pounds	Sent to Klerksdorp for use on the line to Lichtenberg. (REC BW 6)
February 1902 (1st Quarter)	17th (Field) Company RE (Standerton)	1	One "movable blockhouse", pattern not specified, but not circular.	Sent to Klerksdorp for use on the line to Lichtenberg. (REC BW 6)
February 1902 (1st Quarter)	26th (Field) Company RE (Pretoria)	6	Six mobile blockhouses built, type not specified.	Three to Klerksdorp; three to Mafeking (REC BW 9: District CRE Report)
April-May 1902	CRE Newcastle District	2	"Defensible ox-wagons being built"	Northern Natal Quarterly précis of Staff Diaries. (REC BW 2 CRE)
	Total built (recorded)	26+	Up to a maximum of 30 might have been built	

Table of known mobile blockhouses constructed.

CHAPTER 7

LIFE IN THE BLOCKHOUSES

Few, forgotten and lonely,
Where the empty metals shine –
No, not combatants-only
Details guarding the line.

'BRIDGE-GUARD IN THE KARROO' BY RUDYARD KIPLING

Virtually every regiment or corps in the British Army sent at least one battalion to South Africa (the 15th (Kings') Hussars and the 4th (Royal Irish) Dragoon Guards were not present).[260] The scale and extent of the war's impact on the life and experience of the Army of the time should not be underestimated. Many thousands of men and their officers served and died in this far land and a tour would have been physically arduous and demanding. For many the deployment lasted the entire three years of the war. Soldiers of the day from the Foot Battalions would have been either employed in the marching columns deployed across hundreds of miles of veld, in the armoured trains or in static defensive duty on the blockhouse lines.

The shift in duty from the long marches, covering many miles day after day, to the blockhouse lines, would have come as welcome relief. Troops with African support would have built their own corrugated blockhouse, as part of troop, company and battalion lines stretching for many miles. Once completed the lines garrisoned approximately 50,000 British troops of all ranks, with each Rice Pattern blockhouse commanded by an NCO and five to six other ranks, or private soldiers.

The initial period of occupation would have been new and for troops away from their military garrison and battalion life would have come as quite a change in their routine. Now the soldiers were in the position to make their blockhouse comfortable and after improving their defensive position as a priority, would have set about making their blockhouse the best it could be. This so-called decoration and 'beautification' became the norm with each blockhouse, often in competition with its neighbour.

Life in their new environment would have quickly developed, after this initial period of work and excitement, into one of boredom. Life for the soldier now became dull and monotonous, especially during the day when the Boers would avoid movement across the wide open veld, preferring the cover of darkness for movement and crossing.

Night-sentry duty therefore became vital to securing the lines and the dummy sentries in many photographs of the time were of little use to deter the marauding Boer commandos! Each blockhouse also had a complement of up to four African soldiers, mainly used for night duty, but they would have been employed in much of the defensive digging required. They were accommodated in a tent close to the blockhouse, but usually outside the protective wire fences.

Extract from a poem titled 'The Native'[261]

We 'ires them n——s by the score,
To 'elp us fight our Brother Boer,
It 'ardly seems the game;
But likely them as does it knows
Wot they're a-doing, I suppose,
It ain't for me to blame.
Basutos, K——s, Fingoes, h'all !
We takes 'em grateful, big or small.
An' in the Block'ouses they stays,
An' sleeps away the blessed days ;
An' then at nights they'll creep
To little sangars in-between,
Where, if they thinks they can't be seen,
They 'as another sleep.
A gay, light 'earted lot is they,
An' this is 'ow they h'earns their pay.
Whenever they may 'ap to wake,
The h'opportunity they'll take
O' firing once or twice;
But when the Brother comes along,
They sings a very diff'rent song,
An' lies as still as mice.
No doubt they feels too young to die,
An' so they let's the beggar by.

Clearly the offensive language is that of the late 19th century common soldier, and completely out of place today, however it encapsulates in a few lines, life in the blockhouse. The Boers are referred to as 'Brothers' (also referred to as Brethren), the implication is that it is strange to be at war against a white enemy, and that the 'natives' had previously been the enemy in British colonial campaigns. Traditionally at this period of late Victorian history, the wars were mainly colonial wars fighting 'darker skinned' opponents. There was however a distinct need for black troops, both to build the blockhouses and in the manning and guarding of the lines, as this freed up British soldiers for mobile column duty. It would seem Tommy Atkins was grateful for either additional assistance or to escape the blockhouse duty for good and that relationships were pretty jovial and light hearted. In general the Africans believed that supporting the British to an expected win would strengthen their position in the new South Africa. The British would seem to generally have treated the black workforce better than the Boer, but after the war ended, Africans were not afforded any better life than previously. The poem notes there are intermediate sangars that were also occupied in order to further strengthen the line, clearly with mixed results when confronted by a Boer commando at night! It goes on to complain of the two shillings a day pay the 'native' soldiers were earning compared to only one shilling for the British Tommy.

Once the whole system of blockhouses and wire lines with flares and spring guns was established, the night sentry was often the first point of alert, which made for a nervous and often dangerous time. Animals would stray onto the wire, or a storm might rattle the tin cans, initiating a cacophony of fire from the blockhouses, the rifle rack might be initiated and bored troops loose off very many rounds, in an excited frenzy. The practice of shooting only along the wire, which was erected in an inter-locking wave or zig-zag pattern, prevented the blockhouses from firing on each other. In later wars this became known as 'friendly fire' or 'blue on blue'.[262] In one such night-time initiation the shots rippled some 100 miles (160km) down the line.[263] Other instances were less dramatic but equally devastating in terms of loss; it could be a dangerous business, being on duty in a blockhouse, bored all day long and scared stiff by the sounds of the night at others. One such story was that of Private John Huggan's death, which unfolded to his relatives through a series of letters. He was a member of the King's Own Scottish Borderers (KOSB)

The local newspaper reported:[264]

We regret to announce that Private John Huggan of the 1st K.O.S.B. who was wounded at Rooidraal, South Africa on 1st January died of his wounds two days later. It is eight years since Private Huggan first entered the army. He served four years in India taking part in the Tirah and Chitral campaigns. He afterwards returned home and became a reservist. Just two years ago he was called on to join the fighting in South Africa, and has seen a great deal of hard service. Deceased has left a widow and one child, for whom and other relatives much sympathy is felt.

HAWICK NEWS 10TH JANUARY 1902
PRIVATE JOHN HUGGAN 1ST K.O.S.B
ANOTHER HAWICK SOLDIER DIES AT
HIS POST'

The initial newspaper article only reveals that he lost his life whilst serving in South Africa. A follow-up article by the paper on 14th February quotes a letter from his Officer Commanding:

I was with your husband half an hour before the accident occurred. You have reason to be proud of him for though suffering great pain he bore it heroically. The doctor told me he never saw a man fight pain so bravely when he died. The whole post was very sorry; for he was very much liked, as a soldier he was excellent, always ready to do his work thoroughly. We buried him just beside the post, and I will get a cross put on his grave, the men of D Company care for the grave, which is kept nice and neat, some flowers having been planted on it.

Clearly some sort of accident had occurred and this was not as a result of enemy action. The final instalment in the matter comes from one of his comrades-in-arms of his blockhouse, who knew much more of the incident. Private James Veitch writes:

...just a few lines to tell you how my beloved chum was laid low. He was in No. 29 blockhouse along with me. He had gone out without the knowledge of the rest, and we were sitting quite contentedly when we were startled by

the sentry firing a shot. We jumped up and ran out. I asked the sentry what he had fired at, and he replied that he had heard screaming down the valley. He thought they were Boers and fired. By this time we had discovered what had occurred. I found that my dear chum had been shot in the small of the back, the bullet coming out of the groin. We saw it was a bad wound, so we bandaged it as well as we could do. By this time the doctor had arrived and he did everything that it was possible to do. We had him removed to the hospital tent where the doctor worked on him night and day, but it was to no avail: He died on the morning of the 3rd at 7.30, after being unconscious for half an hour. He remarked to the doctor just before he was removed from the field that he had himself to blame; so that it was an accident. All our company are deeply afflicted by the loss of our comrade, and send their kindest sympathy and regards for you all at home. We are making his grave very nice and will send a photo of it.

Typically there would have been challenges and passwords for guards to use when moving between blockhouses. Guards would have shouted the 'Halt!... who goes there?' to be given in return the password for safe passage, but for the African guards other more novel ways had to be found. JFC Fuller, in his book *The Last of the Gentleman's Wars,* recounts one such incident regarding his personal safety. On visiting an African contingent dubbed 'The Black Watch' at No.2 Cossack Post close to Jordan Siding, he was nearly shot for humming the wrong melody! It was his duty as an officer to visit these night posts, spread down the railway line. Africans manning the posts in pairs were taught a popular melody of the day to learn before going on duty in order to allow safe passage for 'friendly forces'. However the training process was either not rigorous enough or the memories of the Africans somewhat lacking, and there were many instances such as this where officers were nearly killed. He thought this near miss somewhat suicidal and vowed never to visit at night again! It was also dangerous for the African guards and one, called 'John No 4', did get shot in another incident.

Each blockhouse had a connection to the telegraph system, which provided connection to the outside world and military updates and briefings. This would have been supplemented by regimental and other army gossip, anything for troop entertainment. The soldiers wrote letters home and the close proximity of the rail lines, to some, ensured a steady flow of weekly mail, which is so vital to a soldier's morale. In addition an officer would visit each day to inspect his men and check on their situation, or to bring supplies, such as rations or ammunition.

Rations of the time were meagre and often the blockhouse garden would grow maize and other vegetables and sometimes keep stock such as pigs, goats or chickens for the pot. Normal rations in South Africa were bully beef and hard tack biscuits, occasionally supplemented by fresh meat, eggs and vegetables. The sparse rations might also be made up from a well-established kitchen garden with the occasional pumpkin or ear of corn, or a garden just made for decoration. Many of the larger tin containers were recycled as plant pots or made into trench art to decorate the area. Today at many battle sites and places of encampment, including blockhouse positions, there are still waste bully beef tins lying discarded in the surrounding area.

Bully beef tin near the Witkop blockhouse. A plethora of rusty rubbish can still be found close to the site of any British occupation on the veld.

Perhaps the lack of good rations, in some small way, was recognised by Queen Victoria, as her present to the troops for the New Year 1900 was a tin to each soldier containing chocolate, with the message 'I wish you a happy New Year – Victoria' written on the lid. Most of the chocolate was likely consumed on opening, but some was preserved and survives today. The tins are not rare as tens of thousands were produced by Rowntree and Cadbury for distribution to the troops in theatre.

BOER WAR RATIONS

Even though the Commissariat and Transport Department did supply a certain amount of tinned goods to the unit, much was either obtained via private purchase or, in the case of volunteer units, at the expense of the unit commander. Generally, tin cans were of a commercial quality, and sported labels as would be found in stores. There was a great variety of tinned goods available, but on campaign, the following would normally be encountered in the packs of troops:

1. Tinned meats, fowl and fish (both issue and private purchase varieties) – corned beef, pork loaf, liver paste, sardines (in cylindrical tins rather than the current flat ones), chicken meats and turkey bits.
2. Tinned milk. Both the evaporated and the sweetened condensed varieties were in use, and were greatly prized by a nation of tea drinkers. The most common size was the 12-ounce tin, and many a Pattern 88 bayonet still bears nicks near the tip from frequent use as a milk tin opener!
3. Tinned 'hash' and other ready-to-eat items such as the various stews were also encountered quite frequently, with sizes ranging from 12 ounce tins right up to 48 ounce tins.
4. Tinned beans (in a variety of sauces) were also quite frequently encountered.

When touring Boer War battlefields today, you know that you have found the general vicinity when you come across large quantities of ancient-seeming tin cans, flattened by time and by the soldiers themselves!

During the war, British soldiers were issued with coffee and coffee grinders, as coffee was considered an important part of the evening meal! Of course, the troops were very likely to promptly use the coffee mills to grind mealie meal, which was issued rather coarsely ground as part of the portable ration) into cereal for stewing, rather than using them for their intended purpose.

Taken from: 'From Beef and Chocolate to Daily Ration – British Rations in Transition 1870–1918'.[265]

Boredom and troop morale were probably the greatest problems faced by the military command of these units. Gardening and gambling were favourite pastimes, but some ingenious troopers also spent time devising additional alarms involving old tins, pebbles and trap guns set to go off should anything brush against the wire. This latter activity probably made the wire lines even more sensitive to activation and shooting into the night by jumpy sentries was more prevalent. As Major General JFC Fuller wrote after the war, 'the blockhouses were demoralizing... men could only talk, smoke and gamble, and the monotony of food, no civilization, it was like living in a medieval castle for days on end...Masses of men were bored to extinction.'[266]

In an extract from a regimental story, one soldier recounts: 'This was in spite of the dreary life led in the blockhouses; in each of which seven men lived together in a small space, the only entry to which was a man-hole. There was one officer to every four blockhouses, whose duty it was to keep the small garrisons continually at work improving the defences, keeping the field of fire clear of undergrowth, and repairing the roads and drifts. In addition each man had to walk to the neighbouring blockhouse twice a week. The Colonel rode round continually, to ensure that all this was being done.'[267]

Life in these small blockhouses through the seasons was most likely very different. The corrugated iron walls and roof would act like an oven during the long summer and like a fridge during the short winter months, when there might even be the occasional snow shower. The weather could make life very uncomfortable in a blockhouse, where the daily fluctuations and seasonal variation might cause a fluctuation of 30–40°C. In one day therefore it could be very hot during the day and very cold at night. The 2nd East Kent Regiment (The Buffs) noted[268] that they experienced temperatures of 32–37°C on the Nelspruit to Komatipoort line during the January summer of 1902.

There was no place for a fire, and no internal chimney in the Rice Pattern blockhouses and food was most likely cooked outside, preventing any warmth being provided internally. The need for firewood was very keen and any wood provided for the blockhouse was more likely to be burnt for fuel than used for its intended purpose. There is evidence that troops made their own wooden furniture, such as tables and washstands, to make their lives more comfortable. Of interest in the rare interior photographs is that the need to live inside a blockhouse actually impeded the full and easy use of loopholes from which effective fire had to be made, often at night. The blockhouses were designed to fight from, but not really to live in at the same time.

Officers and Other Ranks (ORs), whilst coming from different social classes, were separate in all ways whilst in the field. Other than sleeping rough on the veld, ORs and officers would sleep separately, as would African guards and ORs. The separation in terms of class and in maintaining social distance in the British Army was carried out largely in the name of discipline. If a junior officer had to spend time close to the men he would invariably find a tent or trench separate to the men to sleep in. Many officers might also come with a horse and their own orderly or batman whose duties included (but were not limited to) acting as the officer's runner to convey orders on the battlefield, maintaining the officer's uniform and equipment, acting as his bodyguard, digging his trench, and other miscellaneous tasks the officer does not have the time or inclination to do! Class hierarchy ensured there was a rank and status to being an officer too. In defensive periods such as this both the Sergeants' and Officers' Messes would have been established somewhere. Where a rum ration was provided to the blockhouse, wine, beer and spirits would have been in evidence in the respective messes. Just as an indication, the young Winston Churchill came to South Africa as a journalist, having not secured a position with a regiment. He left ostensibly to be involved in the war, as a civilian, but one with the social status at least of an officer and member of the gentry. Thus he saw the Boer War expedition to South Africa in 1899 as a jaunt to be supported with various essential provisions. Along with his telescope, field glasses, saddles, and a bronzed needle compass, Churchill took with him the following inventory[269] of alcohol: six bottles of vin d'Ay Sec, eighteen bottles of St-Émilion, six bottles of 'light port', six bottles of French vermouth, eighteen bottles of Scotch whisky ('10 years old'), six bottles of 'Very Old Eau de Vie 1866', and twelve bottles of Rose's cordial lime juice. There is no indication that this was sufficient for the campaign!

Life in the larger masonry blockhouses was probably less demanding and less claustrophobic, as they were occupied by up to 20 men. They had two large water tanks in the basement, an area for storage and a much greater living area. The middle floor was their living space and the upper floor allowed the guard and the sleeping soldiers to be separated. No doubt the guard rotas were less frequent and in general these Key Point blockhouses saw far less action due to their formidable defensive fortification, as the Boers kept their distance. Very few saw any action at all, and the soldiers were probably kept more active and interested in passing trains as a source of entertainment, than as a possible target of any marauding Boer forces.

Two very different views of life in a blockhouse: contemporary photographs of the insides of blockhouses were challenging for the cameras of the day. The picture on the left, courtesy of Angloboerwar.com, is from inside an early Rice Pattern octagonal blockhouse, while the photo on the right, courtesy of King's College London Photograph Collection[270] is from a larger stone-build blockhouse, showing the officer's or senior NCO's bed and washstand on the top floor.

Routine and discipline went hand-in-hand living in a defensive role and a typical set of daily orders for a unit, sourced from a handwritten set of orders from The Queen's Regiment,[271] reveals a daily cycle of checking and cleaning. Many units would cycle through months and years of this type of static duty intermingled with column work and in supporting the Drives.

Orders for Blockhouse all ranks.

1. All drinking water is to be boiled.
2. Never more than two men at a time are to be absent from a Blockhouse unless specifically employed on fatigues.
3. All ranks including fatigue parties will take rifle and bandoliers when leaving a blockhouse beyond the limit of the latrine.
4. All sentries day and night to wear proper khaki with helmet, puttees and at night the bayonet.

NCOs responsible that:
A. The ammo, and rifles and all Blockhouse equipment are correct and should anything become unserviceable report to the Coy Officer at once.

B. That all drinking water is boiled.
C. That the Blockhouse and surrounding area are kept scrupulously clean.
D. That all men shave daily and that hair is kept properly cut.
E. That all rifles are daily inspected.
F. That latrines are kept in perfect order and whenever possible dug six feet deep.
G. That the rifle rests are constantly tested to see that they bear down the line of wire correctly.
H. That all wire entanglements round and between Blockhouses are kept in thorough review and no clothes hung on them.

By order ...... Signed G Whiffing
Bn Major and Adjt of The Queen's Regiment

These low-level orders were most likely provided in the absence of others or to supplement higher-level orders issued by battalions or divisions in the defensive lines. The more extensive example shown here from the King's Own Scottish Borderers goes into much more detail, and includes how to manage the African guards, night patrols and train incidents, etc. The two different sets of orders also show how the regimental system worked, and that the regiment was the highest level to which the soldier looked for guidance. Each literally beat to its own drum and different nature: note Order No 4 might be typical of a Scottish regiment – NO 'Surrender!'

HUNG UP IN ALL BLOCKHOUSES STANDING ORDERS[272]

1. The garrison of this post is...2... N.C.O. and ....8....men.
2. It will be visited daily by an Officer.
3. No Excuse will be valid, that its DEFENCES in all respects are otherwise complete, including obstacles, wire entanglements etc.
4. Its surrender will entail trial by Field General Court Marshal: and the penalty, on conviction of any individual forming part of the Garrison is a penalty to DEATH or penal servitude.
5. The rifles of the Garrison will be attached and secured near each loophole, and the men's accoutrements and bandoliers similarly se-

cured and a reserve of 200 rounds of ammunition per man will be maintained.

6. The water supply will be renewed daily and water tank reservoir duly protected from rifle fire.
7. No cooking or eating of meal is permitted close to the work. This must be carried on outside at least 30 yards from the work.
8. Line watchers are attached to this post. They will be allowed to live in a shelter near the post (a part of the tent if available). They will be posted in groups of 2 to 3 at distances of about 400 to 500 yards along the line from their post and at a different place every night, and move after nightfall. The Commander of the blockhouse will do this.
9. PATROLS from the GARRISON will move by night both North and South of the post to the second group of night watchers posted on the line from the posts, and arrange with the 'natives' a code of WHISTLE SIGNALS (Code to be varied occasionally).
10. Two groups of night watchers, if available will be posted between BLOCKHOUSES.
11. The sentry by day, will be posted on some convenient spot some distance from the blockhouse where he can get a good view of the surrounding country; by night he will be posted in the trench inside the wire entanglement.
12. The rifles and ammunition of the night watchers are to be stored by day in the blockhouse.
13. The use of watch dogs is to be encouraged.
14. The post is furnished with Telephones and Signal Rockets and the visiting Officer will see that the Garrison knows how to use both.
15. Men are not to hang blankets or clothes on barbed wire entanglements. Plain wire fencing can be used for this purpose.
16. Any hostile movements near the Line are to be immediately reported to the nearest post, and by night a Signal Rocket will be fired as well.
17. In foggy or misty weather extra patrolling is to be carried out along the line by day and night.
18. NIGHT FIRING: Should the enemy attack a post or piquet at night and be driven off, the Garrison is to continue firing for at least five minutes; as the sign of the enemy having ceased fire may be taken that they are in retreat.

19. DANGER SIGNALS in case of explosions on the line or accidents to Trains. A man must be sent to the highest point of the gradients both North and South of the danger point. BY DAY he will stop approaching trains by raising both arms extended (if available) raising a red flag. BY NIGHT he will place fog signals on the line.
20. VISITING NIGHT PATROLS will move out at uncertain hours and one visit will be always before a rising moon and after a sinking moon.
21. Night watchers should be issued with 20 rounds of ammunition; and arms and ammunition taken from them by day and stored in the blockhouses to which they are posted.

BY ORDER. J. W. Godfrey Colonel CSO[273]

The British press, represented by *The Times*, was ever keen to portray life in South Africa from their various correspondents. By early 1902 one such newspaper story[274] tells of the British Soldier or 'Tommy Atkins' as he is colloquially referred to as being 'quite settled into blockhouse life'. But life was also dangerous, not just because of Boers or accidents, but also from silent killers. Disease was a common problem and more of the British force in South Africa died of diseases such as dysentery, typhoid (enteric) and intestinal infections (60%) than from battle wounds (35%).[275] The table below shows the figures for all the British deaths during the war, and a slightly higher figure noted in *The Times* reporting of the day. In terms of accidental death 86 men died from lightning strikes, indicating the severe hazard of living in a corrugated iron shed on a mountain! Those blockhouses in the highveld and mountain areas are noted to have some sort of lightning protection system fitted. The records of the day[276] also note that one soldier died from being eaten by a crocodile on the Usutu River! In addition to these deaths there were also another 40,000 men who were wounded during the war.

Cause of Death	Numbers	Percentage
Killed in Action or from wounds	7,582	35%
Disease	13,139	60%
Accidents	1,221	5%
Total	**21,942**	

Causes of death during the war.

Poor sanitary conditions were the primary cause of diseases such as dysentery, typhoid, and intestinal problems, and simple acts such as washing hands were difficult in the scarce water conditions of a blockhouse. Rivers similarly might have been contaminated, and there were hundreds of thousands of cattle, sheep and horses also using the river as drinking water. Each battalion had a medical officer and they would have provided advice in in sanitary their conditions. There would have been a separate toilet or latrine dug, at a distance from the blockhouse, termed at the time as the 'Khazi'.

A train journey from the Cape Colony up country would have passed numerous blockhouses and by this stage the soldiers might have been manning these defensive positions for quite some time. The railway line was fenced in with hundreds of miles of barbed wire in low entanglements strewn out between each blockhouse, guarding the open veld and river beds and mounted on small hills or koppies. These blockhouses had become home in a distant part of the world – but one which the ingenious soldier had made into an often homely and unique dwelling. The soldier's competitive sprit within his own unit and of course between other units meant there was often an attempt to outdo each other.

Typical postcard of the war showing all the comforts of home.

Rows of white-painted stones spelled out unit names, declaring the inhabitants' regiment, or cap badge motifs; often each had a sign up with a comical name, such as 'Kruger's Villa', or 'Rundle's Starving Eight'[277] and 'Chamberlain's Innocent Victims'.[278] Clearly the officers of the time allowed these names to persist, as part of the British sense of humour of the period. The unit names might be 'The Manchesters' or 'The Glosters'. Paths led from the rail line to the blockhouse, if distant from the track, with steps cut into the hill if the ascent was steep. A train trip from the Cape Town to Pretoria is almost a thousand miles (1,600km) and would have passed by very many blockhouses on the way – and to the ever-imaginative soldier the train was a source of both entertainment and supply.

PASTORAL CARE – A CURATE'S TALE[279]

Each regiment or battalion had its own padre or pastor during the war, but many also came out from Britain to supplement these. One such was William Drury, a young curate who came out to be based in the fortified town of Warmbaths from March 1901. He undoubtedly visited the fine example of a square blockhouse that remains in Warmbad today. He had replaced a chaplain who was on sick leave and had no formal training to assume his role over the garrison troops and local military hospital and blockhouse line troops.

He seems to have been quite a character and was known for using a 'railway cycle' to travel up and down the line to visit the troops. It would have to be removed, no doubt with some haste, were there a train competing for the same line (although he also used a horse on many occasions). He was working on a 60 mile (96 kilometres) stretch of blockhouses line to the north and south of Warmbaths. As he described it, he had a parish 60 miles (96 kilometres) long and 4 foot 6 inches wide (1.3 metres).

Drury had a fortnightly schedule and the troops would have seen him once in that time, equipped with only a Bible and some service sheets. Services were usually conducted in the open air and they sang such simple church hymns as 'Abide with me'. The men were most likely happy to have a visitor as it broke up the monotony of life and he felt genuinely appreciated.

Interestingly he notes that places on his line were marked in kilometres, not miles, and each blockhouse was noted as a number from Pretoria; thus

Blockhouse 94 was 94 km from Pretoria. On his railway bike he would take an armed cyclist with him, and enjoyed the ride plummeting down the hill, but fails to say how hard the cycling was back up. He found the zinc blockhouses fine places for singing: the sound of the hymns was excellent when services were inside, although he confesses to having struggled with some of the Yorkshire dialect, proclaiming it 'terrible Riding'. On several of his days he would visit seven or more blockhouses in one day, and at some locations like Groot Nyl Bridge he would find three blockhouses close together and might get something like a congregation of 14 for his services. At Pienaars River bridge he found three stone and three corrugated iron blockhouses. The company he visited here had been in the field for ten months and not fired a shot in action. In addition, in this part of the line they were living rather well and the curate breakfasted on eggs, buttered toast, fried tomatoes, bread with syrup and, toast!

A railway cycle.

Signs would be shown to the trains requesting 'Papers please', at which the passing train windows would open and a flurry of printed paper would respond to the appeal. The 'Tommies' would dash to the line saluting and cheering to collect their papers and often other gifts of cigarettes, chocolate, fruit and anything else the passengers cared to donate.

In many areas the dummies used in the early part of the war had given way to more comical ones, especially where there was now fighting or Boer movement. Pipe smoking dummies, rather plump well-fed ones, or Kitchener look-alikes, all still resplendent in their khaki uniforms and medals, had been made and well kept by their keepers. No doubt some had their own names, perhaps even making it onto the occasional ration roll for resupply in jest! Some were even manually automated and would salute passing trains, to the amusement of all.

TO RELIEVE THE MONOTONY OF LIFE IN A SOUTH AFRICAN BLOCKHOUSE
DRAWN BY W. RALSTON

A comical look at how the soldiers kept themselves busy on blockhouse duty.[280]

The soldiers were portrayed as happy-go-lucky and plucky, and good-humoured, who all made the passengers 'swell with pride'. The blockhouses certainly were now reflecting their resilient ability to make good of their surroundings and living conditions. Discipline seemed to be still in good order – no doubt with some slight relaxations made in terms of the more decorative and less militarily functional nature of the blockhouses. They were after all a temporary structure and once war had ended would be removed quickly, leaving only a shadow or slight scar on the open veld. Many of the sites are clearly discernable; the corrugated iron left circular piles of stones and shingle as a mark of where it had once stood. Surrounding and connecting trenches are also still visible in many parts of the old blockhouse lines.

The light-heartedness of the outward signs of peace in these later accounts also belies the severe nature of the war, and there is mention of small, walled cemeteries containing white crosses. The war had been heavy in casualties, of both those who died in battle or from disease and those transported as wounded back to England. No doubt the soldiers who still resided close to the cemeteries tended the graves of their fallen comrades.

Towards the end of the war, when the attacks were less frequent and the logistics and resupply lines of communication far more secure, life became more luxurious for some troops at least. Private Ernest Griffin of the 3rd Leicestershires, who found himself stationed in a blockhouse between Knapdaar and the Burghersdorp line, tells that the often-reported meagre life was not so meagre after all. He says in a letter home, dated May 1902:[281] 'It is a "toffs life in a blockhouse" no work to do but to sleep and eat. As for the eatables we get very good food, the ration train comes every two days and drops at each blockhouse ¼ of a sheep, 6 loaves, 4 pounders (a 'pounder' was typically one tin of preserved meat, usually bully beef) and 1 pint of rum, then every 12 days a supply train comes with sugar, coffee, tea, jam or marmalade, potatoes, candles, pepper, salt, etc. and don't forget there is plenty of it.' He continues in a reassuring tone, '...we can afford to eat 2 tins of jam a day. Whenever you hear any body calling the army about the neglect of the troops you call them a liar and stand upon me for the proof. Each man is also entitled to 1 pint of beer a day but the regiment has got to get a canteen train up first.' On the subject of guard duty, he goes on to say that 'taking his turn at night only means 1 hour on guard and 4 hours sleep in rotation, and gets 8 hours of sleep each and every night.'

The 4th Battalion The Rifle Brigade, in a letter to the editor of *The Rifle Brigade Chronicles* dated December 1902, again gives a good feel for the life and monotony of blockhouse life.

4th Battalion letter to the editor
Bloemfontein
Orange River Colony.

Our time on the blockhouse line was bound to be somewhat monotonous and tedious; and though it was varied by fairly frequent sniping at night, the novelty soon began to wear off and the erection of a barbed wire fence of wonderful proportions could hardly be termed an exciting task; still, the Battalion can flatter itself that its line of defence was acknowledged to be one of the best in South Africa, even though it was affirmed that the Battalion had permitted the ubiquitous De Wet to pass through our line, a fact which the gentleman in question, after peace was declared, denied, and informed us that, on the date in question, he was at least 70 miles south of our line!!!.

> *So much for the accuracy of some 'Intelligence Reports'.*
>
> *More than one drive terminated either against or close to our line, but unfortunately we gained very little excitement from them, the wily Boer not coming anywhere near us. Sergeant White and his myrmidons built the officers a palatial mess house at Doornkop, and various other smaller houses to shelter Duff's mighty stores, and Sergeant Nash's little knick-nacks. If one did not learn a great deal that was particularly interesting on the blockhouse line, one at all events acquired the art of becoming a first class thief: without this art one cannot exist in South Africa. June 1st brought us the news of peace, and almost directly afterwards we were ordered to send off a party to attend the Coronation festivities at home. Wilson went in charge of our little lot, and they sailed on the Bavarian. Hardly had this party left us when we got orders to move into Kroonstad, Lieutenant-Colonel Leslie taking over duties of Commandant from Wilson. We soon found ourselves established in our old quarters on Gun Hill, and, greatly assisted by the Pioneers under Sergeant White, made ourselves comfortable. We forgot to mention that before leaving the blockhouse line the authorities kindly requested us to roll up all our wire fencing, a task we could not relish, considering all the pains which we had spent upon making our fence impassable.*

Despite the intense boredom being experienced by the troops Lord Kitchener was reporting his success and continuing to convince the British public of the resolve of the 'plucky troops', in the *London Gazette* of 25 March 1902:

> *With the continued extension of these lines, the strain of night duty, and the labour of constructing fences, trenches, and other obstacles, has fallen heavily on the lessening garrisons of individual blockhouses, and it is a pleasure to me to draw attention to the cheerful and willing spirit in which the troops employed upon these isolated lines continue to respond to every fresh call which is made upon them.*

CHAPTER 8

EFFECTIVENESS OF THE BLOCKHOUSE SYSTEM

This policy of the blockhouse might equally well have been called the policy of the blockhead.

GENERAL CHRISTIAAN DE WET, 1902

Was the blockhouse system a war-winning strategy or, as General Christiaan de Wet suggested, a 'policy of the blockhead'?[282] In fact it was not a strategy in itself but part of the wider strategy to defeat the determined and resilient Boer commandos. Only opinions have previously been aired in this regard, and even Pakenham was not in a position to provide objective evidence to quantify the blockhouses as a success or a failure. It is wrong to consider the blockhouse's contribution without the other elements in the total war-winning context. However, when considered against the specific threats already mentioned in Chapter 3, the blockhouse system's effectiveness can be better quantified. This chapter aims to determine the effectiveness of the blockhouse system once and for all.

AGAINST BRIDGE DEMOLITION

After the declaration of war in October 1899 the Boer forces attacked many key points in Natal and the Cape Colony. They penetrated into British territory and took control of key railway infrastructure, which had quickly been identified as of strategic importance as it would afford rapid mobility to troops disembarking from the ports and moving northwards. These bridges, in military terminology, thus became strategic key points, which were required to be defended and covered by defensive fire. As the pendulum swung in favour of the British and the Boers retreated northwards they sought to destroy this infrastructure in order to deny its use to the British. Bridges and rail infrastructure destroyed by the Boer forces as they retreated is shown in the table below. Many of these were long high-level multi-span bridges.

Rail Infrastructure	No. of Bridges	River	State
Norvals Pont bridge	1	Orange River	Cape Colony
Bethulie Bridge	1	Orange River	Cape Colony
Colenso Bridge	1	Tugela	Natal
Standerton Bridge	1	Vaal	Transvaal
Fourteen Streams Bridge	1	Vaal	Cape Colony
Vereeniging Bridge	1	Vaal	Transvaal
Modder River Bridge	1	Modder	S. of Kimberley Bridge
Every Bridge/Culvert north of Bloemfontein	100's	Modder, Sand, Vet, Vals, Doring and Renoster Rivers	OFS
Frere Bridge	1	Bloukransriver	SE of Colenso, Natal
Transvaal Bridges	9 Pretoria to Komatipoort line and Barberton branch line	Various	Transvaal
Cape Colony Bridges	7	Various	Cape Colony
Smaller bridges and culverts	140		OFS and Transvaal
Tunnel at Laings Nek	1		Natal
Miles of damaged track	Unspecified		
Locomotive engines rendered inoperable	222		Transvaal
Rolling stock	4,200 items		Transvaal
Coal destroyed	1,000 tonnes		Transvaal

Railway infrastructure destroyed by Boer forces in 1900.[283]

Major Percy Girouard RE, who had been transferred from Egypt and appointed Director of the South African Railways, was completely taken aback by the sheer scale of destruction, stating that: 'The enemy had destroyed the railway in a manner probably unprecedented in any campaign. The damage in the Orange Free State was so numerous that the work of the construction train was unable to keep pace with the Army, in spite of every effort to expedite the work.'[284] Having learned that the rail system was vulnerable to this level of destruction, the British set about its defence, initially with earthworks and sandbagged trenches and then in a more concerted effort with blockhouses and armoured trains. As they moved northwards, repairing and securing bridges, culverts and stations, they also secured the key points and commenced patrolling the lines. Later in the campaign, once the masonry blockhouses were constructed, starting with those in the Cape Colony and then in the occupied Republics, no further direct acts of sabotage were conducted on these fortified positions. The Boers had lost all their artillery pieces early in the war and the only way to attack these key points was by direct attacks or using stealthy bridge demolitions with explosives, both of which were high-risk options. The strong defences of the blockhouses now made hard targets of rail infrastructure, and ones requiring the risky strategy of large direct attack with no guaranteed outcome. The Boer forces were not prepared to fight the British in this manner, preferring to move out to more open areas and commence their tactics of destroying trains on the move, which were far more vulnerable and easier targets to defeat with less risk involved. In terms of effectiveness therefore the Elliot Wood and then the Rice Pattern blockhouses used to defend rail infrastructure were completely successful, but while the blockhouses were still being laid out to cover the distance between the defended bridges, culverts and stations, the remaining lines and trains continued to be vulnerable to attack. The Boers then became notorious for train attacks, and started to significantly impact the British logistics and war fighting capability.

COUNTERING THE TRAIN ATTACKS

Having already demonstrated their capability to demolish bridges and destroy rolling stock during their offensive phase, the Boer forces now moved onto the open veld. They knew full well that the long distances and amount of materiel required to keep the British moving north would be considerable and this destruction would significantly slow down their progress. The British engineers did repair all the bridges, culverts, and water tanks destroyed and then sought to protect them with fortifications and troops.

The key area for Boer attack and British response was along the Pretoria to Komatipoort lines where train wreckers such as the notorious Jack Hindon wreaked havoc on the British trains. Once Pretoria was taken in June 1900 this line eastwards to the coast became strategically important in opening up a third and much closer line of communication to the port of Lourenço Marques. The Delagoa Bay Line now required protection and was at the forefront of blockhouse consideration for the local Royal Corps of Engineers Company based in Middelburg. After a visit to that area with the company commander, Major Spring Rice, in September 1900, Lord Roberts returned from Komatipoort to his staff in Pretoria with the first record of the term 'blockhouse' mentioned in theatre diaries.[285] Protection of the line followed initially with the use of stony sand to in-fill the space between corrugated-iron walls of rectangular blockhouses, built by Lourenço Marques contractors. Thereafter, in January 1901, the deployment of six rectangular corrugated blockhouses for rail protection was followed by over 50 octagonal and circular mass-produced designs deployed from March to April 1901 by the Royal Engineers Company.[286] Ultimately this 250 mile (400 kilometres) stretch of rail line would have approximately 480 blockhouses defending it and the key installations along its route.[287] It can be shown from the graph of train attacks on this line versus the general scale of blockhouse deployment that the reduction in attacks is directly attributable to the blockhouses and defence on the rail line.

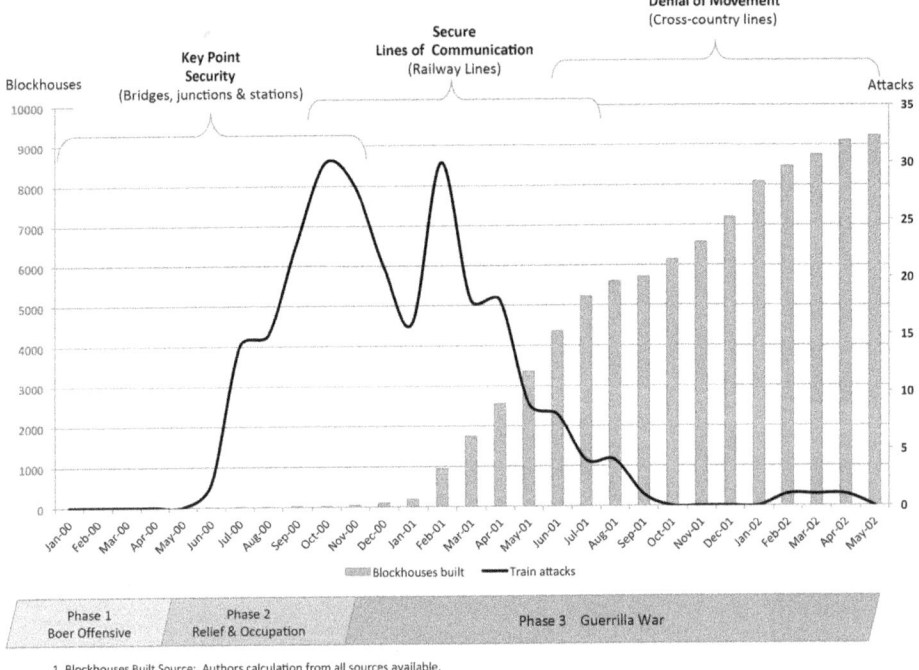

The decline of train attacks in relation to rate of construction of forts, blockhouses and earthworks.

PROTECTION OF THE KEY TOWNS

Early in the campaign Roberts believed that once the key Orange Free State capital of Bloemfontein and the two key cities of the Transvaal Republic, Johannesburg and Pretoria, had fallen, the war would be won. Thus they were protected and defended almost as soon as they were occupied, with an inner ring of town defences and an outer protected area. Blockhouses of all types were utilised and the defences grew as the war dragged on into the guerrilla phase.

The Boers in general were people of the country and not of the city, and when their capitals were captured, they vacated and moved their seats of government and key assets such as their gold reserves out onto the veld. The Boers had no stated strategy to retake these towns, where the British forces were occupying well-prepared defensive positions. The guerrilla phase was characterised by 'hit and run' tactics where they sought to find weakly defended soft targets such as railway infrastructure and small units where they could quickly amass a localised numerical superiority and win through. The towns therefore were not in their plans to attack, although individuals did infiltrate the defences for subterfuge and meetings with relatives, but no major acts of sabotage were conducted. In this regard then the defences were effective but might be considered as only a preventative measure and were not materially tested by the Boers as a potential target.

Johanna Brandt recounts in her book *The Petticoat Commando* that: 'As time went on, Pretoria was being shut in more completely every day. Blockhouses rose on every side; on the hills which lie around the town, and searchlights played from commanding positions over many miles of country, making darkest night as clear as noonday; barbed-wire fences enclosed the entire capital, and outposts were on guard night and day – with no avail!'

Despite this young men frequently 'glided in and out like serpents in the night', often to buy a pound of sweets or a box of cigarettes, or to spend a few days with their friends and to escort their sweethearts to church on Sunday. They also kept those still loyal to the Boer cause informed of what was really happening in the field, in the face of British propaganda. The three towns remained porous to this type of infiltration throughout the war despite blockhouse fortification.

SECURING THE LINES OF COMMUNICATION

The British in the first few months of the war had felt the full might and capability of the Boer hatred for them and their resolve to deny them key strategic assets. In a deliberate act they demolished almost every bridge and culvert on their rail system, and destroyed huge quantities of trains and rolling stock. This placed a huge burden onto the Royal Engineers to repair and build temporary bridges in order for the advance north to continue. Although their lines of communication were to extend, they did recover and rail lines continued to be the main method of strategic supply for the remainder of the war.

The Boer destructive capability was well recognised by Roberts, who early in the campaign made sure that the railways were secured, initially with sandbagged defences and then with extensive blockhouse construction. Many of the key Elliot Wood Pattern bridge blockhouses have survived in the Cape region, although the interconnecting Rice Pattern blockhouses that stretched from the Cape to Pretoria and beyond have long since been removed.

The early destructive might of the Boer exercised on their own unprotected infrastructure was relatively easy to effect, without risk to themselves. Later, when they had lost all their artillery pieces and were scattered in roving commandos, there was a definite reluctance to engage in well-defended masonry blockhouse sites. Consequently there was not one attack on a key bridge, station or siding after their deployment. In addition the 20 armoured trains were also key in adding to the deterrent the Boers faced; although based at permanent sites they could be moved quickly to reinforce against a potential threat. British reviews in this area were unanimous in declaring success, and they were confident that this was due to the blockhouses acting as a key deterrent.

DENIAL OF MOVEMENT

Denial of Boer movement was mainly associated with the building of the cross-country blockhouse lines from the period August 1901 right up until May 1902. Kitchener had resolved to use the blockhouse lines of communication along the railways as barriers against which to 'drive' the Boers in operations similar to a country estate game shoot. The early use of the thinly spread South African Constabulary posts to supplement these had encouraged the deployment of extensive cross-country blockhouse lines to increase the area of operations and effectiveness of the lines for these drives. The concept was to break down the veld into smaller parcels or areas in which to control and ultimately capture or destroy the Boer commandos. Without the blockhouse lines, the drives would have been a continuous and long chase around the veld.

Although Boer commanders such as De Wet with his determination and aggression could pass at will through the lines, his forces suffered losses he could not afford to take each time this had to be done. The blockhouse lines were still somewhat porous in this regard but many other Boer commanders did fear engaging blockhouse lines and where possible avoided crossing them. To that end they were at the very least a frustration and relatively effective obstacle. Once fully effective, with deep counter-mobility defence ditches, it was reported that that men and horses could pass through, but that every crossing lost some of the ox-wagon supporting echelon.

Each blockhouse line came with its own telegraph line, the eyes and ears of its occupants adding observation over long corridors of veld that were used to direct large mobile columns to engage with commandos in a more directed manner.

As a hard barrier against which to drive Boer forces, blockhouses were less effective than they were at preventing movement. The lines along the railways had the most success, as they had two sets of wire fences, one either side of the track, and could also call on one of the armoured trains should a commando approach be detected, but the cross-country lines, in more remote and hilly terrain, provided more opportunity in terms of 'dead-ground' to effect a stealthy crossing. All attempts recorded at crossing the lines were during darkness, although some had to be achieved during a full moon. There is evidence that the African night observation posts and the British blockhouses were at times reluctant to engage Boer forces during these crossings. Where trips were initiated and 'automated fire' from the fixed mounted rifle racks fired down the wires, these were one-off fires, and once completed, racks had to be reloaded manually. These inherent weaknesses were no doubt noted and taken advantage of by Boers who did not want to suffer any unnecessary losses.

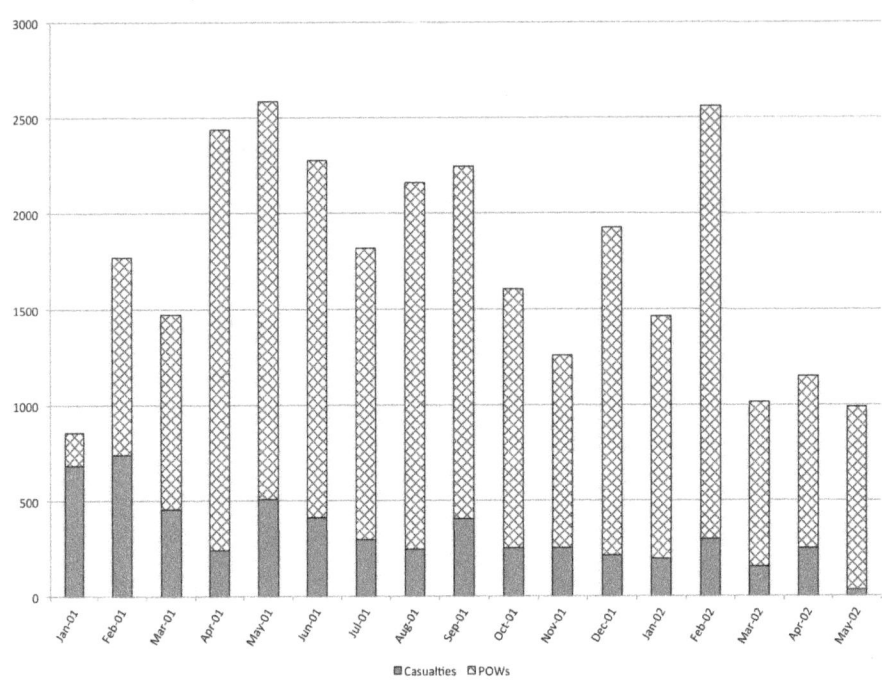

Boer POW/surrender and casualty figures during the 'drives' phase of operations, January 1901 to May 1902,[288] taken from the 'Blue Books'[289] presented to Parliament in 1906.

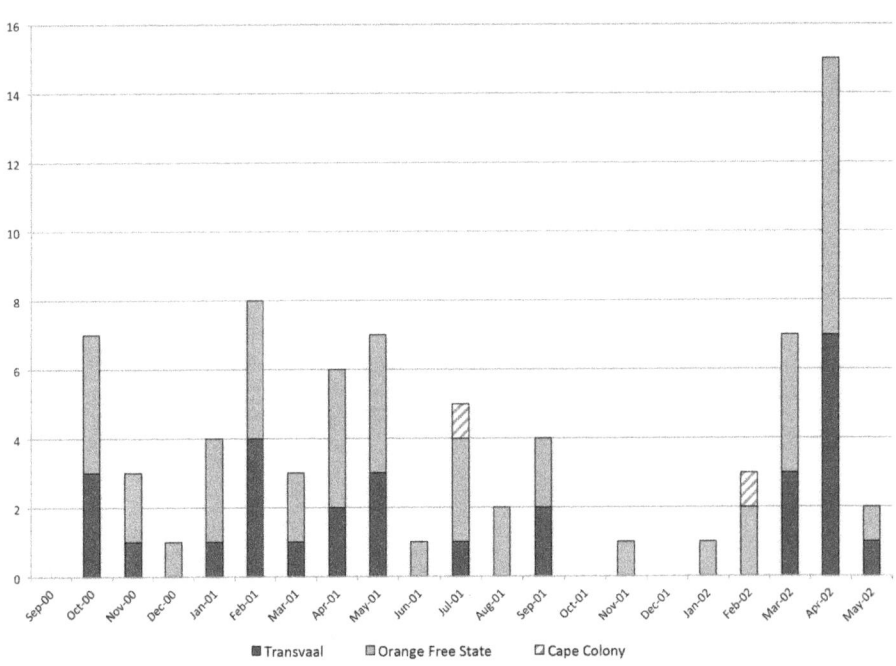

Total drives by area, September 1900 to May 1902.[290]

EFFECTIVENESS OF THE BLOCKHOUSE SYSTEM

The effectiveness of the armoured train combination with the blockhouse system as a means of defending the railway and denying movement was proved on 27 June 1901, when the Boer forces under General Ben Viljoen were surprised while crossing the line. In the action that followed Viljoen's forces suffered heavy losses: six men and 40 horses and mules killed, then a further 25 men killed or wounded, 35 cattle killed and a wagon destroyed by a shell from a 12-pounder gun on board the armoured train.[291]

There is also evidence that during the campaign there was a continuous development of the capability of the blockhouse lines. Wire fences were initially weak in terms of their anchorage; stays were easy to remove and large lengths easily laid flat by pulling on the picket posts. As the operation continued these were improved upon and dug in much deeper, and alarms added to detect early tampering by the Boers. Fences that could not be pulled down were cut using conventional wire cutters, which prompted the wires to be either braided into thicker wires or strengthened with annealed wire shipped out from Britain.

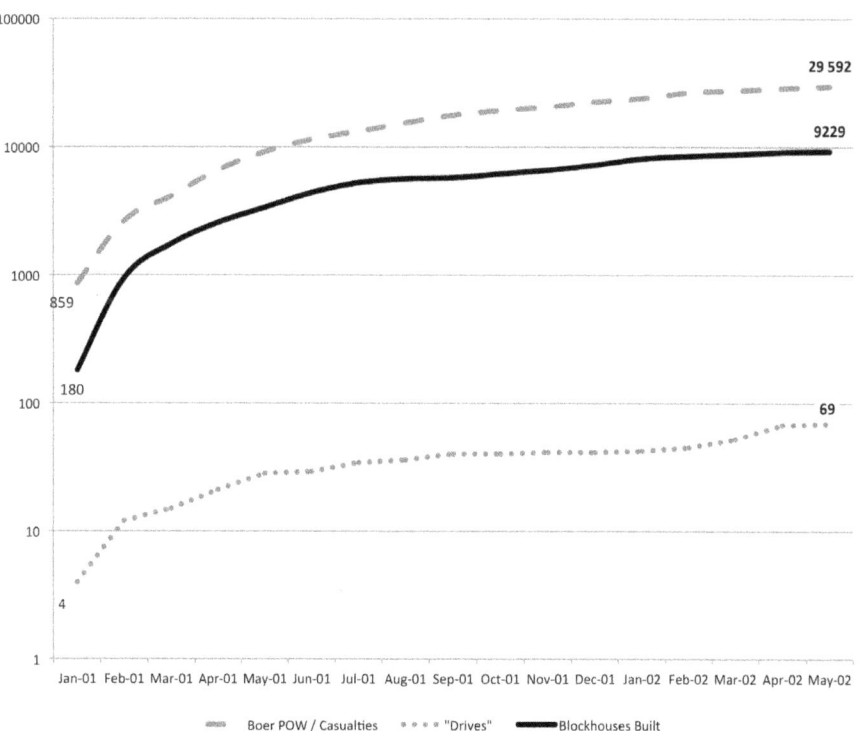

Rate of degradation of the Boer forces against the rate of 'drives' and blockhouse deployment on a logarithmic scale.

The graph shows on a logarithmic scale how the two-pronged approach of 'drives' and blockhouses achieved success in defeating the Boer commandos. Over a 17-month period the British deployed over 9,000 blockhouses (See Appendix A) and conducted an average of four drives per month (total of 69), resulting in the reduction of Boer forces by nearly 30,000[292] (1,700 per month captured or killed). By May 1902 the Boers only had 21,000 burghers[293] left in the field. It had pretty much been a gradual and continual war of attrition that had worn down the Boers to such a level that they were no longer combat effective. Although not the sole reason for the British achieving final success, these two key strategic deployments of engineering resources and mounted columns ultimately proved their worth. Other key factors that should not be ignored were the lack of Boer logistic support in terms of arms, ammunition and horses, the overwhelming superiority in number of the British forces, an increasing fear over the numbers of black and coloured people fighting for the British (by the end of the war some 140,000), and the plight and psychological impact of the civilians in the camps.

COST EFFECTIVENESS

The total cost of the war was approximately £210 million.[294] Exact figures are difficult to determine because the costs were borne by many sources. For example, the colonial units were initially funded by their source governments; however, the figures that follow, below give a reasonable estimation.

Date	Cost
1899 - 1900	£ 23,000,000
1900 - 1901	£ 63,737,000
1901 - 1902	£ 67,670,000
1902 - 1903	£ 47,500,000
	£ 201,907,000
Interest added	£ 9,249,000
Total	**£ 211,156,000**

The breakdown of the cost shows how the money was spent in the major categories of the components of the war effort. 'Works, telegraph and engineering works' would have been the bulk of the costs relating to the Royal Engineer 'Works, telegraph and engineering' tasks, and the blockhouse system would have been within this £4.7 million.

Item	Cost
Supplies	£ 47,600,000
Ships and transport	£ 30,500,000
Other stores	£ 17,470,000
Horses, mules & oxen	£ 16,525,000
Railway costs	£ 15,700,000
Pay for regular forces	£ 14,500,000
Clothing	£ 9,400,000
Wages for transportation staff	£ 7,670,000
Pay for South African forces	£ 7,500,000
Pay for Imperial Yeomanry	£ 5,150,000
Works, telegraph and engineering works	**£ 4,700,000**
Compensation	£ 4,580,000
Ammunition	£ 4,315,000
Pay for Militia	£ 4,000,000
Concentration camp maintenance	£ 3,540,000
Gratuities for troops	£ 3,500,000
Pay for overseas colonials	£ 2,700,000
Pay for medical establishments	£ 2,270,000
Pensions (to 31 Mar 03)	£ 1,660,000
Miscellaneous costs	£ 1,270,000

The author's calculations, shown in Appendix B, suggest that the total material costs for the system of over 9,000 blockhouses and 6,000 miles (9,600 kilometres) of line (excluding labour cost of over 80,000 British troops and up to 36,000 Africans) would have been in the order of £1.6m (somewhat higher than the figure suggested by Bethell[295] in his report to be 'well under £1,000,000'). It might well be that the true expense of the blockhouses lines was deliberately understated. It is of note that the Royal Commission held on the war later in 1902 failed to raise any question concerning blockhouses. Major General Elliot Wood appeared before the Commission, but the only major questions raised were concerning the lack of light railway and pontoon bridges available in theatre.

Even with a calculated increase in cost, in terms of the total bill for the campaign the cost of 'blockhousing' even at £2 million only represents a mere one per cent and could be considered outstanding value for the British pound spent versus the contribution it gave. From another perspective however, it was the bloodiest, longest and most expensive war Britain engaged in between 1815 and 1915. It cost more than £210 million and more than 22,000 men were lost to Britain. The Boers lost over 34,000 people and more than 20,000 black, Indian and coloured participants were killed.

THE BOER VIEW

After having taken time to inspect a blockhouse line, riding up the line about ten miles (16 kilometres) late in December 1901, Lord Kitchener was told the story[296] of a pithy Boer who had described the line's effectiveness by saying that 'if one's hat blew over the line anywhere between Ermelo and Standerton one had to walk round Ermelo to fetch it'. Indeed many of the Boer burghers did fear crossing the wire and were wary of engaging these well-prepared lines of fortified blockhouses.

There was a reluctance and wariness against direct engagement during the latter stages of the war, although some members of the Boer leadership, such as General Christiaan de Wet, were openly dismissive of the blockhouse policy, describing it as 'the policy of the blockhead'. Certain lines, such as the one from Bloemfontein to Ladybrand, were singled out for specific mention, being called 'a White Elephant' – useless, troublesome and expensive. In his post-war memoirs[297] he poured scorn on the British for the tactics used and wondered why 'England the all-powerful – could not catch the Boers without the aid of these blockhouses'.[298] In retrospect he was far more fearful of mobile troops encircling him at night in his laager than the blockhouse lines. Had the British adopted these tactics, he declared, the war might have ended sooner and not been prolonged by at least three months building more blockhouse lines. Others, like General Louis Botha, were more realistic in their assessment when they stated during the last days of the war: 'A year ago there were no blockhouses. We could cross and recross the country as we wished, and harass the enemy at every turn. But now things wear a very different aspect. We can pass the blockhouses by night indeed, but never by day. They are likely to prove the ruin of our commandos.'[299]

De Wet had proved time and again that a well-led, motivated and prepared commando could pass through the blockhouse lines. Often the breach points were well reconnoitred and intelligence gained on their disposition, night-time sentry routines, and areas of dead-ground. The Boers planned their breaching operations meticulously whenever possible. Often this was very close to blockhouses and observation posts, and since they had forewarned the African population that any local caught supporting the British would be killed, many were very wary of even being seen, let alone engaging the forces as they crossed the wire. In actual fact their duties were only to act as a human trip wire and alert the blockhouses to an attempted crossing. Even when challenged and fired upon at close range, De Wet recounts cutting the wire and passing through safely with no casualties. Despite the numerous 'drives' and 'New Model Drives' he was never captured, however his son Jacobus 'Kootie' was captured during February 1902.[300] It was largely due to his stubbornness that De Wet senior failed to credit the British with any success with regard to the blockhouse system.

General JC Smuts, later Prime Minister of the Union of South Africa, summarised the British system of waging war in a report made in January 1902 as follows:

Lord Kitchener has begun to carry out a policy in both (Boer) Republics of unbelievable barbarism and gruesomeness which violates the most elementary principles of the international rules of war ... The basic principle behind Lord Kitchener's tactics had been to win, not so much through direct operations against fighting commandos, but rather indirectly by bringing the pressure of war against defenceless women and children.[301]

In his view, the blockhouse lines and drives were not so devastating as the scorched earth policy targeting the civilian support infrastructure. If there was an impartial view from the Boer side regarding the blockhouses it was most credibly written by Denys Reitz in his book *Commando*. His account and description at every turn and event is factual and largely impartial. He concludes that the blockhouse system played its part in the gradual reduction in combat effectiveness of the Boers still fighting the war as follows:

I know little of the actual Peace Conference as I was not a delegate, but the outcome was a foregone conclusion. Every representative had the same disastrous tale to tell of starvation, lack of ammunition, horses, and clothing, and of how the great blockhouse system was strangling their efforts to carry on the war. Added to this was the heavy death toll among the women and children, of whom twenty-five thousand had already died in the concentration camps, and the universal ruin that had overtaken the country. Every homestead was burned, all crops and livestock destroyed, and there was nothing left but to bow to the inevitable.

THE BRITISH VIEW

Arthur Conan Doyle[302] considered the blockhouse system to be both 'a elaborate and wonderful system' and 'the chief factor ... in bringing the Boers to their knees'. The truth was most likely not so exact as this, but as part of a wider strategy this was more or less the case. There are very few documented cases of attacks or any significant action at blockhouses although in the British official records there are a few cases mentioned.

In a letter in late 1901, Ian Hamilton, Kitchener's Chief-of-Staff, reporting back to Lord Roberts, states that: 'every single individual I have spoken to from General to Subaltern, seems to think that by this method (of blockhouses) we shall most rapidly bring the business to a close.'[303] There was no dissent in the chain-of-command regarding the tactics of blockhouse lines, and the concept was widely regarded as both innovative and part of a more technologically advanced approach to the warfare and defeating the Boers.

De Wet's conviction that the blockhouses were a 'Blockhead Policy' must have rankled with those involved. It prompted Elliot Wood to write to *The Times*[304] on his return to the UK, protesting that, 'certainly as far as the attacks on the trains were concerned, the blockhouse protection had virtually stopped all acts of sabotage'. He went on to state that the blockhouses limited the Boers to having to cross the lines at night, already corroborated by General Botha, and that British convoys subsequently had much safer routes to follow along the blockhouse lines, and that they provided good observation of vast parts of the country to gather enemy intelligence.

1899 NOVEMBER 25
LIST OF RECORDED CONFLICTS AT BLOCKHOUSES IN THE CAMPAIGN

The list of officially recorded actions contains five instances of Boer attacks on blockhouse lines as follows:

1901

26 JUNE

Attack, blockhouse, Delagoa Bay railway line, Transvaal by General Viljoen who captured a seven-man blockhouse wounding most of the defenders.[305]

7 AUGUST

Incident, Brandfort: blockhouse line, Orange Free State; attack Vanrhynsdorp, Cape Colony

24 OCTOBER

Engagement, Kleinfontein (Driefontein), Marico; Attack Blockhouse line, Badfontein, Transvaal. Lt Mellor RE with his blockhousing parties on the Lydenburg Road were heavily attacked, but the escort of the Devons etc easily beat off the enemy with some loss.[306]

26 DECEMBER

Engagement: Senekal, Blockhouse line, Hanover Station, Cape Colony

1902

19-24 JANUARY

Attack, concentration camp/blockhouse line, Pietersburg, Transvaal

After the war Major-General Elliot Wood had appeared in October 1903 at the Royal Commission on the War in South Africa[307] to give evidence and answer questions regarding the provision of engineering support to the war effort. The minutes reflect that the key issues highlighted and discussed were the lack of railways and bridging supplies available, the quantity and provisioning of stores, and other experimental equipment used such as steam traction engines and motorcar searchlights. In 21 pages of evidence there is not a single mention of blockhouses and the effort required to construct over 9,000 of them, suggesting that this was conducted well within 'existing capabilities'. Only the sourcing of barbed wire was mentioned, and that stores had to be procured locally in order to meet the demand. Other witnesses at the Commission did mention blockhouses but only in passing as it related to their field of operation. In medical terms and in the fight against disease, blockhouses were useful in that daily life allowed for the boiling of water and regular provision of food, and provided a healthier lifestyle than that in a roving column on the veld. One of the killer diseases was enteric (or typhoid) which was carried in dirty water and was responsible for a high casualty rate amongst British, Boers and Africans alike.

Maps shown below published in *The Times* newspaper sought to show how progress was being made in defeating the Boers. Actually what they highlight is that the more the Boers were harassed and squeezed in the eastern Transvaal, the more they were pushed out into roving into the western Cape region. Attrition was having its full effect and their combat effectiveness was gradually becoming weaker. When Deneys Reitz travelled with General Smuts from O'OKiep in the far north-western Cape until shortly before the Peace Treaty at Pretoria in the Transvaal, the one stark observation he made in contrast to the two forces was that the Transvaalers were in a very poor state of health and were all but finished.[308]

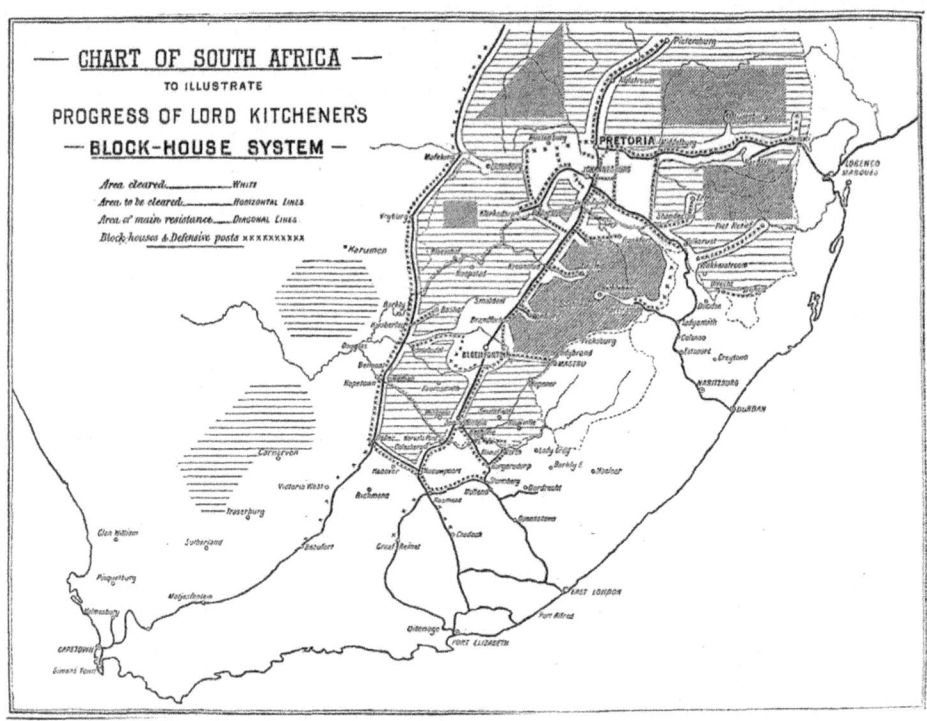

Map in The Times dated 25 December 1901.

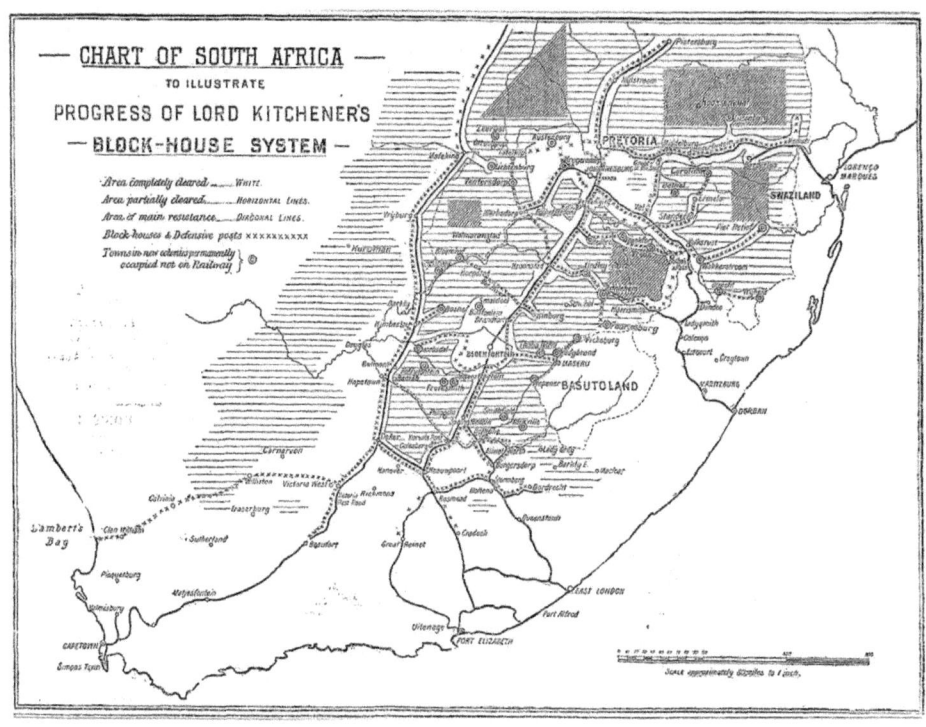

Map in The Times dated 24 January 1902.

CHAPTER 9

THE ARCHITECTS OF THE BLOCKHOUSE SYSTEM

To provide meaningful architecture is not to parody history but to articulate it.

DANIEL LIBESKIND

Who takes credit for the blockhouse system and the design of the small forts and blockhouse buildings in their plethora of sizes and shapes? Ultimately the system was credited to Lord Kitchener, as it was started in earnest to counter the guerrilla phase for which he was the commander-in-chief. Several, however, claim to have had an influence, or an outright claim that it was their idea, and this chapter explores all those to whom some of the credit might be given at either a strategic, tactical or technical level.

STRATEGIC ARCHITECTS

Lord Roberts

Frederick Sleigh Roberts, 1st Earl Roberts, c1897.[309]
Courtesy of the National Portrait Gallery, London.

Along with Wolseley, Roberts was a great general of the British Empire: steadfastly they promoted it and along the way made sure their own careers went from strength to strength. Resolutely in competition with each other, they formed extensive and competing circles or 'rings' of competent, loyal and influential officers whose careers tended to follow their leader across the Empire. Wolseley had an African or 'Ashanti Ring', while Roberts's ring was the 'Indian Ring'.

Born to an Army family while serving in India, he came home in 1834 to be educated at Eton in 1845, and then entered the Royal Military Academy (Woolwich), gaining his commission in the Royal Artillery in 1851, and was posted to the Bengal Army. So commenced an early career based in the far-flung but influential part of the British Empire. India was to be a productive place for the young Roberts to establish himself. By 1858, he had his first and only battle wound, for which he earned his first of seven mentions in dispatches fighting at the time of the Indian Mutiny. Near the village of Khudaganj he achieved the ultimate accolade, winning the Victoria Cross.

Aged 28 he was already promoted to brevet major, married and, after a period of ill health convalescing in England, back in India serving on the North West Frontier (1863) in the Quartermaster General's Department. In this capacity he no doubt learned the skills to sustain troops in the field and appreciate the importance of planning, not a skill every fighting general of the period possessed, and which would stand him in good stead years later advancing to Pretoria.

He continued to suffer from periods of ill health, but service continued in Abyssinia (1868) and the Lushai Campaign (1871–1872). It was here that he first led troops into battle and was quickly promoted to substantive major in 1872. His rise through the ranks was swift and he quickly established a reputation during the continuing Afghan campaign, leading the Kabul Field force to first retake Kabul and then relieve Kandahar. It was for this epic, but perhaps over-rated march that he became known as 'Roberts of Kandahar', giving him almost celebrity status with the British public.

During the very brief First Anglo-Boer War he was dispatched to be the new Governor in Natal and the British Force Commander-in-Chief as the war was not going well. Unfortunately for Roberts the war came to a quick and unfavourable end for the British while he was still on the boat travelling to South Africa. This left him with a feeling that the Boers were 'unfinished business', especially after the defeat at Majuba and the easy terms on which the British agreed peace, which also left the Boers believing that Britain had little heart to fight for South Africa.

Nineteen years later, the three crippling defeats of 'Black Week' and the ever-despondent telegrams from Buller had the Cabinet and Whitehall back in London wondering if they had the right man for the job. Buller was also pondering if he was strong enough to resolve the situation and relieve Ladysmith. The determined man who had faced the Zulu spears and won a Victoria Cross all those years before, had now mentally and physically gone to seed, largely through a good appetite for food and fine champagne. Queen Victoria was his most celebrated admirer but now even she thought that he might abandon Ladysmith. The Empire was teetering on the brink of disaster, the beginning of its end.

Roberts was in his Ireland headquarters, keenly following the events as they unfolded. His sense of loyalty and devotion to his beloved Empire would not allow him to sit back and let matters unfold. He should have lobbied the Commander-in-Chief, but this was Wolseley, his rival, and that avenue would not have proved successful. In true strategic military fashion, he chose another route and lobbied directly to Lord Landsdowne, the Secretary of State for War, and an old ally from India. Gradually winding up the case for himself as the right man for the job, his letters variously discussed the question of replacement in case of incapacitation of the generals, or that the force was too large for one man, or finally that the British might have to sue for peace in light of the current failures! Roberts was beginning to get London's attention.

He had nothing left to gain in terms of reputation: his success in Afghanistan had topped an outstanding career; conversely, he had everything to lose in South Africa, but how could the situation get any worse? Balfour, the Foreign Minister of the time, confirmed his appointment as Commander in South Africa, with Kitchener as his Chief of Staff. A formidable command team was now ready to depart for Africa. London's nerves were settled and the public had a hero to send off to save the day, much as Roberts had been sent to do previously to avenge the British defeat at Majuba in 1881. For Roberts this might be seen as taking on where he had left off, finishing a job that had been half completed with an ignominious peace that had lasted only a few years. In the meantime, he had learnt of his only son's death at Colenso, a bitter blow to him and his family. Externally he bore the news stoically but internally he surely had a renewed resolve to settle matters with the Boers once and for all, so that his son had not died in vain. He completed his meeting with the

Queen and was bid a hero's farewell at Waterloo Station before sailing with Kitchener from Southampton on the SS *Dunottar Castle* on 21 December 1899. For Queen and Country, two of the Empire's finest generals left to quickly win the war, or so they believed.

En route they concocted their strategic plan, a feint in the area of Norvals Pont to draw the Boers' attention, followed by a main attack in the area further to the west at Orange River Station. The plan was to get the Boers off balance, outflank them and then strike for Bloemfontein and Pretoria. It was a huge Army Group to lead and manoeuvre, the likes of which had not been deployed before in the Empire's smaller wars. Roberts had not commanded in the field for some time and Kitchener, although battle-hardened from the Sudan, was not accustomed to an enemy so well equipped and prepared. They were however going to arrive with considerable reinforcements and would have huge numbers of troops and materiel with which to work.

On their arrival they quickly set about trying to repair some of the weaknesses already exposed and exploited by the Boers. At the tactical level the Army lacked mobility and was short of horses and men to fulfil that role. Shooting, or musketry as it was still referred to, was still executed by firing in massed volley under the command of officers; the importance of range and sight-setting were also not appreciated. Movement in the field and fieldcraft were still not fully appreciated although the Rifle Brigade on the Tugela Heights later in the campaign executed the first use of fire and manoeuvre. Movement and attacks were largely in prearranged and predefined close or open order formations, which made easy targets for experienced Boer marksmen. Soon after his arrival Roberts issued instructions to try and remedy this situation, urging careful reconnaissance, more movement in open order formation to disperse troops, more delegation down to battalion and company commanders, and most importantly of all to make better use of the ground for cover from fire. For greater mobility, three new mounted regiments were raised, two from the Cape, Roberts's and Kitchener's Horse, and from the South Africans, Brabant's Horse was formed. It would leave the Cape vulnerable and under-protected as Milner feared, but Roberts did not fear a Boer invasion. In addition two divisions had to raise and equip a mounted company from each battalion, which all greatly added to the counter-mobility required to fight the Boers.

The railway lines, which were so important for supply, were also a limiting factor for military strategists. The British were unconditionally tethered to the railways; Roberts, on the other hand, wanted to spring a strategic initiative and surprise the Boers. To undertake this he had to get away from the railways, even though this meant a march of over 500 miles (800 kilometres) from the Cape Colony border to Pretoria – no small operation. The transport system was reorganised to accommodate this, with the majority of troops now brought into a centralised organisational structure and command, which was not universally popular. This worked relatively well until the conventional phase had run its course, when it was broken down again to accommodate the 90 mobile columns then used to hunt down the Boer guerrillas. It took Roberts and Kitchener only five months to take Bloemfontein and Pretoria, and even Buller moving at his snail's pace joined up with Roberts one month later in July 1901.

Colonel Bethell, in his official post-war papers of 1904, cites Lord Roberts as the man to first use the term 'blockhouse' in September 1900 after he returned from a visit to Komatipoort. At this period train attacks were continuing to rise and no doubt Lord Roberts, commissioned as an Artillery officer, would have met and listened to advice from his Engineer staff officers. Bethell, a commissioned Engineer, confessed: 'Exactly what it was intended to mean then I do not know...'[310] Roberts however was a veteran from the Second Afghan War and in 1881 as Commander-in-Chief of the Madras Army was noted for improving the defences of the North-West Frontier,[311] which consisted of fortifications and blockhouses.

Although Roberts had started the punitive farm burning as acts of reprisal, gained from his experience of using it on the North-West Frontier of India, he disliked it and knew full well its residual impact. He told the House of Lords that it would result in a 'rich harvest of hatred and revenge'.[312] At least he foresaw the day when Boer and British would be part of the same British Empire serving shoulder to shoulder. This marked Roberts, compared to Kitchener, as a more compassionate and politically astute general and strategist; others however just viewed Roberts as being too soft. It was Kitchener who would embody the image of the ruthless British general, stern and imposing, into the next era of South African politics, and the Boers would continue to hate the British for years to come.

Roberts returned to Britain in December 1900 to assume command as C-in-C of the British Army aged 68 years. He was finally appointed Wolseley's successor at Horse Guards in Whitehall, with a £100,000 reward from Parliament for his services. The British Army was continuing to learn key lessons in South Africa and the Army needed to be reformed as it came out of the Victorian era. Despite being considered 'old school' due to his age, he went as far as he could in modernising the army, bringing in better officer training, more realistic troop training, and procuring better technology such as the new Lee-Enfield rifle and motor transport. When the position of Commander-in-Chief was abolished after the Royal Commission in 1903, Roberts went on to serve on various influential committees. When the First World War broke out and his beloved Indian Army Corps was deployed to France, he resolved to visit them in the field. Unfortunately this was to be his undoing, as the ill health that had plagued his career finally got the better of him. It was November and he caught a chill which developed into pneumonia, and died on 14 November 1914, in St Omer, France.

He was buried with full state honours and interred in St Paul's Cathedral, next to such great leaders as Nelson and Wellington, and of course his eternal rival Wolseley. He will be forever remembered as 'Bobs' or Roberts of Kandahar, and takes his place in history as the marching general who trekked across vast tracts of land to defeat forces daring to threaten his beloved Empire. Although starting many of the unsavoury policies in South Africa, his were noble efforts; he left Kitchener to 'mop up', and one wonders if he would have prosecuted the operation differently or been as successful as Kitchener.

When Roberts arrived in theatre there was already a stalemate and the under-preparedness and weaknesses of the British Army had already been exposed. As he had done in Afghanistan, he had tipped the scales in favour of the Empire. Strategically he mastered the art of out-manoeuvring his enemy, but in both cases had failed to defeat them in the field. He left too much of the Boer force intact and able to exploit their greatest strength of mobility; to make matters worse, they had vast tracts of veld over which to exercise it. Kitchener now had to defeat the Boers, master counter-mobility and conquer the open veld.

Lord Kitchener

Herbert Horatio Kitchener, named after Admiral Horatio Nelson, was born in 1850 in County Kerry, Ireland. He was educated privately in Switzerland and at the Royal Military Academy (Woolwich), gaining a commission in the Royal Engineers in 1871. His mother died of tuberculosis early while they were in Switzerland leaving Kitchener to become a shy and introspective young boy, not given to making friends easily.[313] Famous as the face of the the First

World War recruiting poster 'Your Country Needs You!' and as one of the last generals in Queen Victoria's reign, he became the single figure to take the blame and vitriol for the inhumane treatment in the concentration camps. As the Commander-in-Chief he was certainly accountable, but how much of this was by his initiative and design, and how much of the blockhouse system also credited to him was of his own concept and making?

Lord Horatio Herbert Kitchener, 1st Earl Kitchener of Khartoum, (c1895).[314] Courtesy of the National Portrait Gallery, London.

Prior to the Second Anglo-Boer War Kitchener's primary claim to fame was the relief of Khartoum to avenge the death of General Gordon and for which he became known as Kitchener of Khartoum. He had fought in many of the typical Victorian 'small wars' against 'native' enemies, after having gained his engineering credentials in postings and as an instructor at Chatham in 1873. His first posting in 1874 was to the Middle East where he served in the Palestine Exploration Fun (PEF) as a surveyor of mapping. He also served in Cyprus and Anatolia, before moving to Egypt to train Egyptian cavalrymen. During his leave he travelled to Bulgaria as an observer of the Russo-Turkish War. He also once undertook a clandestine mission in Egypt, where his eyesight was damaged, leaving him with a permanent slight squint. During this time he also learnt Arabic, adding to the French and German he had already become fluent in from his early life in Switzerland.

When trouble broke out in Egypt in 1882 Kitchener saw action there as a captain serving under Wolseley, after which he transferred to the newly British reformed Egyptian Army. His language skills however soon put him to effective use working in intelligence on the staff and in leading small bands of Sudanese irregulars in the desert.[315] Action in Zanzibar came next in 1885, and in Suakin on the Red Sea, where he first gained command experience and was again wounded, returning to England for treatment and recuperation.[316] His reputation also came to the notice of Queen Victoria during this period and his climb up the ranks continued steadily but with little fanfare.

On promotion to colonel in 1888, and after a short stint as ADC to Queen Victoria, he took command of the 1st Sudanese Brigade, later commanding the Egyptian Police, and in 1892 became Sirdar (or Commander in Chief) of the Egyptian Army. During this time he finally avenged the death of General Gordon by defeating the Dervishes at Omdurman on 1 September 1898 and ensured British supremacy in the area by facing down the French at Fashoda. It was to date his finest hour, and he was promoted to Governor-General of the Sudan in 1899, but by December 1899 he was ordered to accompany Roberts to South Africa as both a national if not Empire hero and Roberts's Chief-of-Staff.

The issue of the death of Major-General Gordon of Khartoum and the Battle of Omdurman is interesting in that it sets the scene for the next conflict in South Africa. Gordon was sent to Khartoum in early 1884 to secure the evacuation of soldiers after a local revolt by the Mahdi, Mohammed Al-Sayd Abdullah. However, he became trapped in the city after refusing to leave and was subsequently killed and 7,000 defenders slaughtered by the Mahdi's forces. If this was not disgrace enough for the Empire, the Mahdi had Gordon's head cut off and presented to him. Kitchener had been part of the relief force sent to relieve the siege in Khartoum, but it arrived too late to save Gordon. Once Khartoum was retaken Kitchener found a personal letter addressed to him by Gordon. It was Gordon's last communication and he took the death badly, feeling a duty to be the avenger.

The Battle of Omdurman in 1898, 13 years later, was the culmination of months and months of detailed planning, moving the Army across the desert on a specially constructed narrow gauge railway 370 miles (600 kilometres) long from Wadi Halfa to Atbara Camp.[317] This was a typical use of an Engineering asset to bring almost 1 million square miles of the Sudan back under Anglo-Egyptian control. Commentators after the war termed the building of this railway 'the deadliest weapon ever used against Mahdism.'[318] Perhaps as a former engineer Kitchener appreciated the need to conquer the desert, as it was foremost in his planning[319] that the South African veld would need to be tamed and dominated in a similar manner. Once Khartoum was taken he then decided to destroy the

Mahdi's huge tomb, but first had his body dug up, the head cut off and the torso thrown into the Nile. This was to revenge the death of his friend Gordon who had suffered a similar fate at the hands of the Mahdi. It shows a barbaric and callous nature to Kitchener. Queen Victoria was 'Not Amused' when she learned of this desecration.[320] There was still more controversy after the alleged neglect of the enemy wounded left in the desert.[321] Despite these horrific actions he was lauded on his return to Britain in 1898 as a national hero having restored British pride and rule in the region. It was the start of a long national affection for the man known as Kitchener; his celebrity status was sealed, at least for the time being.

His reputation as a ruthless leader and meticulous planner[322] had been forged during this key period of conflict in the Sudan, however despite success he was not a confident or particularly capable general. During the victory at Omdurman, the young Winston Churchill, who was under his command, described him as being 'stern, unpitying in spirit and privately blamed him for inhumane slaughter of the wounded'.[323] Kitchener was a Royal Engineer officer until becoming a 'red tab' or full colonel; thereafter he served on the general staff, where he was not as well qualified through experience and study as might be expected of a cavalry or infantry general, and lacked both tactical and strategic knowledge. He did however possess skills; he had a machine-like efficiency as an administrator, and although having a tendency to keep information to himself, had an almost photographic memory. He had gained rapid promotion during his career, and this may have contributed to his occasional lack of self-confidence and tendency to offer his resignation or to find a replacement when situations became overbearing. His lack of social skills denied him much of an Officers' Mess life with brother officers, and he ate and drank in strict moderation, giving him a reputation for being aloof. In contrast to this he had a resolute ambition, and a keen sense of whom to court as aristocratic friends and members of the Government and Foreign Office. On arriving in South Africa one of his keenest critics, Lord Milner,[324] described him as a 'strong self-willed man in a hurry' whose nature was to 'gain his ends by tortuous means' which he credited to his oriental background.

Roberts's period in command passed quickly, and although Kitchener took command of actions at the Battle of Paardeberg in February 1900 he did not cover himself in glory as a field commander. He had a tendency to take complete control and bypass the chain-of-command, issuing orders directly to the units in contact. The victory was not as decisive as it might have been and his mishandling of the situation did not go unnoticed. He took over at the end of the conventional phase of war, once all the former republics' capitals at least had been secured.

Kitchener's was a lonely command early in 1901. Roberts had departed taking men with whom he had shared the burden of command, staff of a like mind and loyal to him. While Roberts returned to a hero's welcome, Kitchener was left with a situation going badly wrong, having to request additional mounted troops to take the fight into the mobile guerrilla phase. Alone and with no chief-of-staff, he faced a lonely and difficult command; he was petulant and moody. Roberts did all he could to support him as the new Commander-in-Chief back in Whitehall, keenly aware of the difficulties he faced.

During the first six months of the guerrilla phase he suffered from isolation, confiding only by letter and telegram with Roberts and Brodrick, the Secretary of State for War back in London. He obsessed over small details and was focused on the task in hand, pushing his staff for detail, which was reflected in these dispatches. None of his staff understood his strategy, however, and there was no one he felt he could take into his confidence. He shouldered the full weight of command and responsibility, and was determined to see the war through to a successful outcome; anything less than this might have caused fractures in the globally spread Empire. Milner was pressing for nothing less than outright victory and in other parts of the Empire tensions were rising with the Russians and Chinese.

The scorched earth policy for which Lord Kitchener was to become so famous, and so vilified – was the key to stripping the veld of its supporting infrastructure. Although militarily effective in theatre it created a political outcry back in Britain. It was condemned by Churchill as a policy of 'hateful folly'[325] and by other senior politicians as 'methods of barbarism'.[326] It was a strategy that certainly played a part in winning the war but ultimately lost the peace in terms of Anglo-Boer relations well into the next century. Ultimately the complete devastation of a nation's population led directly to the Afrikaner nationalism that sparked true Apartheid into life in the late 1940s. Burning farms and interning mothers and children had attacked the core of the Afrikaners' culture and their roots.

Kitchener was certainly frustrated with the Boers' determination to remain in the field and the difficulty he faced in commanding and denying vast areas of the veld. Boer commandos retained the initiative and they could still fight where and when they wished. He wrote several letters back to Britain on this topic and lamented the ease with which they were capable of evading capture while his Army was suffering from a lack of horses and mounted troops with which to catch them in counter-mobility operations. Milner had a counter plan. Whilst on leave in May 1901 he reported that Kitchener's 'drive' tactics were not producing success, which gained some credence with the government, who were impatient for victory. He proposed faster-moving columns with which to pursue the Boers, abandoning farm burning and drives.

Under pressure to reduce troop numbers and reduce the vast expenditure in the midst of a guerrilla phase when he needed more, not less, Kitchener grew ever more frustrated. Throughout the remainder of 1901, there were failed drives and few successes; this was further compounded late in the year by the loss of Benson's Column, one of his best assets. Full of self-doubt and remorse for the death of Benson, he wrote to Roberts, suggesting that if there was a better man for the job then Roberts ought to replace him. The papers were also declaring that Kitchener was 'not much of a strategist', a sentiment with which Kitchener acquiesced in letters to Roberts,[327] and now, a year after Roberts's return, the situation was little better.

During this period in 1901, it could be said there were a series of disjointed and uncoordinated policies taking place, which had yet to come together and take effect. The blockhouses were being deployed to protect the lines of communication; scorched earth and camps were being used to wipe the veld of commando support, and large columns were failing to reduce the combat effectiveness of the Boer units still fighting. Despite his stern and robust demeanour, he was forever reacting with emotion to situations; his fondness for 'swatting flies with a sledgehammer'[328] identified him as a potential risk to the successful outcome of the war. As the war dragged on, his dispatches to London lamented the minor defeats of his columns while seeking to have greater numbers of troops deployed to catch the Boers, which the government found disconcerting. He had his own self-doubt: in a telegram[329] to Lord Roberts late in 1901 he 'wonders if he is to be relieved of command due to the remarks in the English papers last week and also considers if a new commander might do some thing more than I can to hasten end of war'. It was thought Kitchener might be losing his grip; he had also written to Brodrick 'am anxious to get to India, can you help?'[330] Clearly his heart and mind were yearning to get back to his Indian normality.

Recognising the huge pressure now on Kitchener to succeed, and his isolation, Brodrick sent him a General Chief-of-Staff in the form of his best staff officer, Colonel Ian Hamilton. It was not before time and would also allow some concession to Milner's criticisms as well as providing a third party view of Kitchener's strategic ability and the concern over his mental health and physical wellbeing. It was Hamilton's secret task to spy on his new commander and report back to London. The Hamilton Letters to Roberts are still available to read in their original form at the Liddell Hart Centre in London[331] and provide an interesting perspective on both Kitchener the man and the dilemma in which Hamilton was placed. For Kitchener's part, he uncharacteristically welcomed Hamilton with open arms, took him into his confidence and delegated much of the detailed work to him: soon Kitchener was off visiting troops in the field.

Hamilton's arrival was pivotal for Kitchener and the war effort, Roberts was sending a clear message to his former protégé, acknowledging he had been under-resourced, and that help was on hand. Hamilton, despite some apprehension, recognised that he was also there to provide much needed moral support to an isolated and somewhat shy general. He wrote to Roberts on the *SS Britain* on the way down to South Africa in November 1901,[332] that his initial three aims were to:

'1. Cheer up Lord K, and decentralise command structure.' [In order to share the burden – he needed to delegate. This would also allow Kitchener more time to think of higher military problems, ie to strategise!]

'2. Arrange for a month's armistice.' [In order to buy time.]

'3. Initiate rifle shooting practice!' [Clearly he had recognised the need to improve the under-performing British rifleman.]

Despite his 'secret task' to report on Kitchener, Hamilton to his credit offered all his letters to him for review, which the General declined, fully acknowledging his loyalty. Hamilton also saw fit to report openly and honestly on a number of the commanders leading the deployed roving columns, many of which must have sunk promising careers on their return home.

At the end of 1901, in the back-and-forth chases of the hunter and the hunted, there came more bad news to spoil Kitchener's Christmas. The blockhouse lines were now being extended across country, but at Groenkop on Christmas Eve a large British deployment used to protect blockhouse building was overwhelmed. The next day, however, in a trap designed to capture Schalk Burgher and the Transvaal government, the British caught General Viljoen and his commando. A bittersweet success, as Viljoen was sympathetic to negotiating a peace, but had now lost his influence having been captured.[333]

Also at the beginning of 1902 came the Boer victory over Lord Methuen's column in March 1902 at the Battle of Tweebosch, with over 700 officers and men either casualties or captured. Kitchener on hearing the news took to his bed for 48 hours, his nerves having 'gone to pieces'. It was to be the last Boer victory, but had clearly unsettled the Commander-in-Chief quite severely. He never took bad news well, often leaving his staff to deal with the situation.

The lack of strategic planning and the haphazard nature of the policies being adopted now seemed with the arrival of Hamilton to start to gain positive traction in the field. He wrote to London that within six months the highveld would be clear of Boers and the war over, unless they escaped to the vast open expanse of the Cape. This was perhaps a strong motivation for the 350 mile (560 kilometre) blockhouse line from Lambert's Bay to Carnarvon strung out late in the war and completed by May 1902. Blockhouse lines were now well developed and divided the land into more manageable areas. Although not impregnable, they allowed Kitchener to utilise the more mobile and greater numbers of columns to conduct his 'New Model Drives', particularly against the illusive De Wet.

As the war dragged into 1902 there were still setbacks and bad news, and again the temperamental General allowed his sensitive emotions to get the better of him, but finally emerged from his room this time to a hearty breakfast, and a resolve to continue.[334] The tactics that had started under Roberts were however systematically wearing down the bitterenders. The countryside was bereft of support, portioned into areas using the fortified blockhouse lines; columns were continuously hunting the commandos, and there was better intelligence and counter-mobility tactics gained from using the National Scouts.

In bringing the war to an end, Kitchener had secured South Africa for the Empire – but at a staggering cost of over £200 million and a huge loss of life. He had managed to maintain his nerve despite having being close to cracking, and signed the Peace Treaty of Vereeniging in Pretoria on 31 May 1902 with the British now having to make good their destruction in reparations. Kitchener was promoted to General and later Field Marshall and after South Africa became the Commander-in-Chief India. By 1914 he was the Secretary of State for War, and as the only serving British Army Officer to serve in government, he was famous for the recruiting poster and his foresighted New Armies. He was drowned when *HMS Hampshire,* taking him to a diplomatic meeting in Russia during the First World War, struck a German mine off the Orkney Islands, Scotland, on 6 June 1916, with a hold reputedly full of gold bullion worth $10 million.[335] His body was officially 'lost at sea', but he has an tomb in St. Paul's Cathedral, London.

Lord Baden-Powell

Lord Baden-Powell has his claim to fame already: the Scouting movement was conceptualised during his period in South Africa. However, Bethell also states in his post-war report that after September 1900 and before Roberts left the theatre in December 1900, General Baden-Powell 'proposed a scheme for defending the railways by posts of ten men placed one mile apart along the lines'.[336]

His designs, according to Bethell, were not best suited to permanent occupation, and were only low sunken dug-out earthworks, with no roofs and little in terms of protective cover from fire. He also points out that the Boers still had field guns at this time, which would have rendered them ineffective as protection against such bombardment. The concept was very similar to that adopted in terms of the posts suggested, but they were not blockhouses.

Lieutenant-Colonel Robert Baden-Powell (c1896).[337]
Courtesy of the National Portrait Gallery, London.

Mr Fry of Pretoria

There was a citizen of Pretoria, one Mr Edward H Fry, who also claimed responsibility for the concept of the blockhouse system, having suggested the idea to Lord Roberts in a letter dated 23 November 1900. Edward Fry was an Englishman who had been resident in the Transvaal since 1874, and had performed commando duty in 1876–1877 against the Sekhukhune, and then served in the Transvaal Rangers, registering as a Transvaal burgher. He noted in his letter that he had taken no part in the war to date, and went on to make detailed suggestions based on his military knowledge, service in the field and understanding of the Boer and the terrain in which they operated, as follows:[338]

- *That if a series of forts or military posts were established, say with a radius of 25 or 50 miles round all the principal towns (the distance to be that found most suitable), the country within that area could be easily pacified. The forts, where possible, not to be further than five miles from each other, so that in case of emergency, one could assist the other. The garrison of each fort to consist of 100 mounted men, colonial by preference, with a reserve force at some spot near the centre to proceed to the assistance of any part that may be necessary. Boers can then be driven outside the forts, and the constant patrolling of the garrison will prevent any bodies of the enemy returning, or, if they should succeed in breaking through, which is doubtful, the reserve force will meet them and they would thus be placed between two fires.*
- *When the country within the area of the forts is cleared, a further portion should be enclosed in a similar way and cleared as above, until the enemy submit or is driven into large bodies, when they could be dealt with in the ordinary way. It is unlikely that more than a double line of forts would be necessary as the area is increased, although it may be necessary to construct more than a double line. If so, then those that were at first constructed should be abandoned.*
- *The country by this means would be gradually and systematically cleared, and it would be probably not long before the whole was pacified, as protection could then be given to those that are willing to submit, whilst, at the same time, they could be kept under strict supervision. Cultivation would be resumed, and the refugees could return. The moral effect also on the Boers, when they see the country occupied, will be very great.*
- *The construction of the forts need not be of an expensive character, the walls could be made of sandbags or simple earthworks, with flanking bastions, and a good ditch outside, the area to be only sufficient to afford sleeping accommodation for the men and cattle, and shelter if they should be suddenly attacked. A patrol should start at dawn each day to see that the country is clear.*
- *In fixing the sites of the forts, water is of great importance, Often, however this would be difficult to obtain at the most promising site for the fort. This difficulty, however, could, to a great extent, be overcome by means of wells or wagons fitted with galvanized-iron tanks (of a capacity of 100 gallons each) to carry water, and, as the garrison of each fort would be small, the supply should not be difficult.*

- *The force required to carry out the above suggestions would be moderate. Supposing that an area of country was enclosed by a line of forts extending on all four sides 50 miles – 200 miles, and that a fort was erected at every five miles, there would then be forty forts with 100 men each 4000, and, in addition, the reserve force only twenty-five miles distant from any threatened part.*
- *The area of country so protected would be 2,500 square miles, sufficient for a considerable population.*

Lord Roberts did receive the letter, or at least his staff did, and Fry received a note in reply acknowledging the fact and thanking him for his suggestions. The reply was sent from the *SS Canada* whilst Roberts was en route to Great Britain on his way home and is dated 19 December 1900. Kitchener was already in command and had most likely never seen the letter from Mr Fry.

The persistent Mr Fry wrote once again, this time to the High Commissioner in Cape Town, amplifying his previous proposal, in a letter dated 7 January 1901.[339]

> *I would therefore, suggest that the Boer method of fighting Kaffirs (sic) be followed, but on a larger scale, viz:*
>
> *That on any portion of the country being cleared, a cordon of forts or military posts be formed on the outskirts, to prevent the enemy returning, and I would further suggest that if a cordon of such forts be constructed round the towns, the mines, and on each side of the railways, at such distance therefrom (viz., the towns, mines, and railways) as is found practicable, say from ten to twenty miles, according to circumstances, the area inside this cordon could then be permanently cleared and settled with reliable men....*[340]

Mr Fry went on in this letter to outline in greater detail how the fort would be built and the active patrolling conducted against the Boers in 'clearing operations'. The forts were extensive in size and well spaced with compounds to accommodate Boers who surrendered. Although this was similar in concept to what Kitchener would do, in detail and execution it was fundamentally different. In an almost identically worded response he was again afforded the briefest of replies thanking him again for his suggestions. Clearly Fry was also witnessing the fortification of Pretoria from June 1900 onwards and the early use of redoubts and blockhouses to defend the city, but before they were deployed in numbers.

Another ten months then goes by, presumably with Mr Fry still living in Pretoria, as he writes from the same address and would now have seen the extensive blockhouse system covering the Pretoria environs. The blockhouses were reported on in all the newspapers and by July 1901 all the rail lines were secured and the cross-country line had begun construction to form the pens into which the Boers were to be driven. Fry must have thought his idea stolen, as he decided to write to The Right Hon. Joseph Chamberlain, Secretary for the Colonies, at Downing Street. He enclosed copies of the letters previously sent to Roberts and the High Commissioner and proceeded to use the term 'blockhouse' referring to his 'small forts', which Kitchener has already deployed by the thousand by now. He claimed credit for the ideas of forts, especially those along the railway lines, and also that his protective compounds for the burghers were adopted as the concentration camps. With hindsight he might have chosen to leave the latter out, but concluded that he had been hard done by:

As I consider the system of Blockhouses, which has proved of great utility, originated from my letters, and that they have tended to assist the conquest of the country, and to curtail the enormous weekly expenditure, I have applied to His Excellency, the High Commissioner and to Lord Roberts for some acknowledgement to that effect, which I trust that I shall be considered to have merited.[341]

Finally in December 1902, seven months after the war had ended, he wrote to Chamberlain again requesting recognition for 'his' ideas and asserting that their adoption had led to 'the successful termination of the war within a few months of their adoption. I thus claim to have rendered important service to the Empire, but my claim has been treated with contumely.'[342] Finally he went public on 13 March 1903 with a letter in the *Transvaal Critic*, but to this day has never got the credit he thought he deserved, but now has his place in history!

THE TACTICAL DESIGNERS

Major-General Elliot Wood

Elliott Wood, the son of Doctor Miles Astman Wood of Ledbury, was born on 5th May 1844, and was schooled at King's College School and later went to King's College London. The Wood family originally came from Gloucestershire (formerly of Preston Court) where they have a family tomb in St Mary's Churchyard, Newent, Gloucestershire.[343] His father was described as having an extensive medical practice in Hereford and noted for being the first to introduce the use of the stethoscope in the town.[344] In 1861 he entered the Royal Military Academy and was commissioned into the Royal Engineers in 1864, and then joined the School of Military Engineering at Chatham.[345] It was Elliot Wood who gave his name to the most significant structural legacy of the Anglo-Boer War – the Elliot Wood Pattern Blockhouse, unique for their steel machioullie galleries on the top-floor at diagonal corners. It is unclear exactly how many of the original design were built, or how many were copied, or modified in some way, but today almost 30 have survived in various forms over 120 years after they were built.

Major-General Elliot Wood.
Courtesy of Royal Engineers Museum, Library and Archives.

Whilst serving in Chatham he acquired a love for water sports, including yachting, rowing and canoeing, which he put to good purpose on his first posting to South Africa between 1868 and 1873. During this tour he designed and built the Zephyr Canoe, nicknamed 'Wood's mad idea' by his colleagues,[346] which he proceeded to paddle down the Berg River from Wellington.[347] His quarters while living in South Africa were the Castle in Cape Town, where he no doubt studied the design of the fortifications for future use. On returning to England he was mainly concerned with building barracks and fortifications as ADC to the Inspector General of Fortifications, the War Office, 1880; and for a period also served on Malta, another very fortified island.

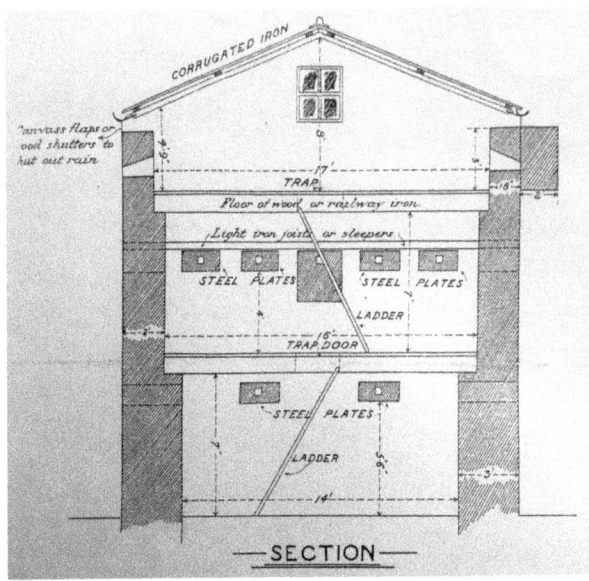

Plans of the Elliot Wood Pattern Blockhouse, which claimed fame in the report on the blockhouse system by Colonel Bethell in 1904.[348]

He first saw action during his posting to Egypt in 1882 and was present at the battle of Tel-el-Kebir, strategically placed between the Suez Canal and Cairo, after which he was mentioned in dispatches and gained the brevet of Major. In addition he was granted the Medjidie Order 4th Class[349] and the Khedive's Star Medal (Bronze Star).[350]

In the Sudan Expedition campaigns of 1884 and 1885 he was mentioned in dispatches three more times, received his brevet of lieutenant-colonel and was made a Companion of the Most Honourable Order of the Bath (CB). In addition he was awarded two more clasps to his Egypt Medal for Suakin 1885 and a clasp to his Khedive's Star. He led the 17th (Field) Company RE from Cairo during the Suakin Expedition and was responsible for planning a railway line from Suakin to Berber.[351] It was during his service here that he developed notable fortifications, forts and blockhouses. Of this experience and after several of the trains he had been riding on were blown up, he writes in his memoirs:[352] 'It was clearly impossible to protect the railways by large posts at long intervals, for in twelve months, as recorded on my office chart, there were no less than 230 "blowings up" of trains or rails. Remembering how at Suakin (where labour was abundant) even two-storeyed blockhouses had been erected in the day, and knowing the enemy had no artillery, I had started to sandbag blockhouses, with steel-plated doors and loopholes in the Komatipoort district...'

The West Redoubt at Suakin (left), courtesy of the National Army Museum, London, compared to a similar design the P'tit Blockhouse (right) in the city of Edmundston, Canada, circa 1835.[353]

One of the interesting examples of fortifications Wood recounts in his memoirs[354] is the one called West Redoubt shown above. Built in 1885, this blockhouse resembles many of the timber colonial blockhouses built by the British in Canada. The P'tit Blockhouse noted above, in comparison, was built around 1835 as part of a line of defences during the Aroostook Wars near Edmundston in Canada and of special significance for the offset second floor. This similar design was incorporated by Wood into the West Redoubt which was 'a sixteen foot square sandbagged ground floor with walls three feet thick, with the second floor placed diagonally across the lower one so that through the floor of the four projecting corners rifles could command the surrounding ditch'.[355] Woods notes that it was built by a party of Sappers and 'natives' in a 'trifle over eight hours', and occupied the same night; on the following night an action against it took place. The soldiers had nicknamed the structure 'Sandbag Redoubt'. Although the feature of the complete upper floor offset did not pull through completely into the Elliot Wood Pattern blockhouse built in South Africa, to some extent the machiciolated galleries at two of the opposite corners did allow over watch and covering fire onto the lower walls, reflecting back to a design from over 60 years before.

After gaining his staff qualification Wood was then appointed Assistant

Adjutant-General at Headquarters and later served in Malta from 1884 to 1889 as Commander Royal Engineers (CRE) Malta. He was quite the designer: a canoe pontoon and superstructure adapted for pack transport was also adopted from designs by Colonel (Sir) Elliott Wood, CB, RE. The pontoon consists of four sections laced together, each section being a framework of wood covered with waterproof sheeting. Three pontoons and eight composite planks form a 'unit', from which can be constructed 48 feet (15 metres) of bridge for infantry in file, 84 feet (26 metres) for infantry in single file, or a raft to carry 15 men or an empty wagon.[356]

Whilst a colonel on the staff (commanding RE) in Aldershot, 1899 he was posted to South Africa to become Engineer-in-Chief serving under General Buller with a local rank of major-general, but was not evidently Buller's first choice, with some military politics no doubt at play.[357] Here he became the chief architect of the initial masonry blockhouses, drawing on his past military study and operational experience. These were to a pattern similar to that he had used in the Sudan during the 1880s. Examples of his blockhouse may still be found at Wellington, Burgersdorp, Harrismith, Wolseley, and Laingsburg and many other sites around South Africa.

In 1901 he served under Kitchener, and it was at this time he developed a system of blockhouses, to ensure railroad security, of which he had some experience. His obituary alludes to difficulties in implementing the blockhouses, stating that he 'had to overcome a deal of official opposition until he was allowed by Lord Kitchener to have his way'. No doubt at some point the idea must have appealed to the latent Royal Engineer in Kitchener. This criticism was also forthcoming from the Boer forces, with General De Wet describing the concept as a 'Blockhead Policy... prolonging the war for at least three months'.[358]

In February 1901, as a Colonel (local Major-General) he was mentioned in dispatches for the supervision of the 'arrangements for pushing forward trenches towards the enemy's laager at Paardeburg ... the successful result attained there is due in a large measure to that Officer's effort'.[359]

Later in April of the same year a despatch from Earl Roberts, KG GCB, to the Right Honourable the Secretary of State for War stated that: 'In continuation of my despatch No 9, dated Johannesburg, 15th Nov, 1900, I have the honour to bring to your notice the excellent work done during the campaign up to the 29th November, 1900, by the various Departments of the Army which have contributed so much to the success of the operations in the field; and mentions by name, Colonel E Wood, CB, RE, as my CRE has given me every satisfaction.'[360]

> *DE WET AND THE BLOCKHOUSE LINES.*
>
> TO THE EDITOR OF THE TIMES.
>
> Sir,—It will doubtless have created some surprise that De Wet says that the blockhouses were of little account. But what are the facts? Prior to the introduction of the system of "blockhousing" railway lines there was a "blow-up" of either trains or railway at some point or other almost every day, with great loss and inconvenience to us and gain to the Boers. When the blockhouse system on the railways was nearly complete there was no such occurrence for months at a time, and the substantial profit on our side was enormous.
>
> It is hardly necessary to go into details or to quote the opinion of other Boers in order to prove the value of the later cross-country lines also. As a matter of fact the Boers, with very rare exceptions, did not attempt to cross these blockhouse lines by day. In many cases they were successfully driven on to them and were captured in considerable numbers. By means of them convoys could safely travel through the country, our columns could find support on them, and the movements of the enemy were far more easily observed and reported.
>
> ELLIOTT WOOD, Major-General.
>
> Aldershot, Dec. 6.

Wood was also very supportive of Major Rice's new low-cost design for blockhouses, especially as their deployment at short regular intervals was having an impact in reducing train and rail attacks. In describing their set-up he also describes the rifle racks and flares system, which was also used in Suakin. During the initial deployment of the Rice Pattern kits from the 23rd (Field) Company RE factory in Middelburg, it was noted by Kitchener on his return to Headquarters after a field visit that[361] 'All the Generals and Commandants on the railways are complaining of this order about little blockhouses: they say they will be mopped up in detail.' After Wood stood his ground with Kitchener, which must have been interesting to say the least, Kitchener quips 'Very well carry on, and you can deal with the protests!'[362] with the manufacturing work then carrying on apace: the Middleburg factory alone produced over 1,000 blockhouses by the end of the war.[363] Citing from the official history, Wood notes with pride that 'the Boers only took out seven blockhouses in the whole war out of the thousands that sprang up over the vast field of operations.'[364]

In April 1901 he was appointed to be an Ordinary Member of the Military Division of the Second Class, or Knights Commanders of the said Most Honourable Order (KCB), for services in South Africa up to the 29th November, 1900.[365] On 29th July 1902 he was again mentioned in despatches by Lord Kitchener:

'I have the honour to submit a final despatch which covers the period between the signature of the terms of peace at Pretoria and my departure from Cape Town. In this I have endeavoured to acknowledge some small portion of the kind assistance or loyal service rendered me by civilians and soldiers during my command in South Africa. Colonel (local Major-General) E Wood, KCB RE, has been indefatigable in rendering me all the assistance in his power.'[366]

Finally for his services in South Africa, on 31st October 1902 Colonel Sir Elliott Wood, KCB was made a Companion of the Distinguished Service Order.[367]

In summary, for this campaign; Major-General Sir Elliot Wood was made a substantive Major-General, awarded the Queen's South Africa Medal (Five Clasps) and the King's South Africa Medal (two Clasps), mentioned in despatches three times and made a KCB – quite a notable achievement to date in his career.

Between 1902 and 1906 he was Commander RE 1st Corps, and in 1906 he was automatically retired from the Army at age 62 years. He married Annie Beatrice Bourne, the daughter of Lieutenant Colonel Robert Bourne and widow of Mr HR Dugmore, and retired to live in Holmer Court, Herefordshire.

Wood was one of the privileged few to own an early camera and was a prolific photographer wherever he went. The Durham University Archives contain two large volumes of his photographs from his service in Suakin, Egypt and during the Anglo-Boer War. Many of the photographs used in CRE reports and post-war papers are photographs taken by Elliot Wood and are of considerable quality for the period.

He again offered his services when the Great War broke out, which were not accepted, despite his notable field fortification experience. Instead he enlisted as a private in the Hereford Voluntary Motor Corps. In 1924 he published a memoir called *Life and Adventure in Peace and War*,[368] which is an interesting account of engineer work (sapping) along the Modder River. Like many of the retired officers of the day, he took up a number of local and county positions such as County Commissioner for Hertfordshire Boy Scouts, County Deputy Lieutenant, a magistrate, vice-president of the British Legion, and vice-president of the city football club. He died at his home in Holmer Park, Herefordshire on 7th September 1931, aged 87 years.

Major Spring Rice RE

Spring Robert Rice was born in July 1858[369] in Tipperary, Knockgrafton, Ireland, son of Spring Rice of Marlhill, County Tipperary.[370] He entered service in the British Army in 1877 where, after his training as a Gentleman Cadet, he gained a commission in the Royal Engineers, at the Royal Military Academy, being promoted to lieutenant[371] on 9th October 1878 and then to captain[372] in April 1888. From 1892–1895, he served in the staff service as a captain in the Submarine Mining Battalion and adjutant of the School of Military Engineering,[373] and at the age of 38 years he was promoted to the rank of major in May 1896.[374] It was while he was at the School of Military Engineering that the Anglo-Boer War broke out.

General Sir Spring Robert Rice (1915).[375]
Courtesy of the National Portrait Gallery, London.

The Times notes that he was posted out to South Africa 'for special service' on Saturday July 8th 1899, embarking from Southampton. The *Cape Times*[376] reported that the *Gaul* arrived on Sunday afternoon with 182 men of the 23rd (Field) Company RE, which he commanded. The Company proceeded directly on to Natal, Major SR Rice in command, with his expected complement of three troop commanders, Lieutenants EV Turner and HL Meyrick, and 2nd Lieutenant RJT Digby-Jones. During October 1899 he was present during operations in Natal, including the action at Lombard's Kop.

The Battle of Lombard's Kop together with one at Nicholson's Nek on 30th October 1899 saw British defeats around Ladysmith that came to be known as 'Mournful Monday', or the Battle of Ladysmith.[377] After a somewhat disastrous attempt to attack Boer forces on these hills, and a catalogue of tactical errors, the British 7th and 8th Brigades were soundly defeated. The British were allowed to retreat by the Boer Commander, General Piet Joubert, leading to the start of the Siege of Ladysmith. The next day General Sir Redvers Bullers, the commander of a powerful expeditionary force, started to disembark his troops at Cape Town, for the start of the relief of the sieges of Ladysmith, Mafeking and Kimberley.

For the duration of the siege Major Rice was appointed Commander Royal Engineers (CRE) and 23rd (Field) Company RE was continuously employed in constructing and strengthening the works of defence. During the defence of Ladysmith he was mentioned in dispatches by Sir Redvers Buller. The *London Gazette* citation read as follows: 'Major SR Rice, RE, acting as Commanding Royal Engineer, has been indefatigable in the discharge of his duties, and his services have been most valuable in preparing the entrenched positions occupied by the garrison of Ladysmith, and in other matters connected with this particular branch.'[378]

The original prototype octagonal blockhouse Major Spring Rice designed (left) before refining the design into the more mass-produced Rice Pattern Blockhouse (right). This octagonal one stood in Orange Grove, Johannesburg for many years after the war until urbanisation consumed it in the 1970s. Courtesy of South African Military History Museum, Ditsong.

Sir George White, in his Ladysmith despatch, dated December 2, 1899, brings to notice the following as being 'eminently deserving of reward:'[379] Major S Rice, RE, acting as CRE, has been indefatigable in the discharge of his duties, and his services have been most valuable in preparing the entrenched positions occupied by the garrison, and in other matters connected with this particular branch. He had shown considerable skill in laying out defensive works and in giving advice and assistance in their construction during the siege.'

In 1900 as Commanding Royal Engineer 4th Division he was present during the actions at Belfast and Lyndenburg, where he was again mentioned in dispatches by Sir George White. As Commanding Officer 23rd (Field) Company RE and while commanding this Company at Middelburg he invented, with the ideas of Captain George Fowke RE, the corrugated-iron blockhouse, to become known as the 'Rice Pattern Blockhouse'.

Major SR Rice RE (seated in the centre) and 23rd (Field) Company RE.[380]

Towards the latter part of the South African campaign he was promoted to the rank of brevet lieutenant colonel, to take effect on 29th November 1900.[381] For the remainder of his tour he was appointed as Commander Royal Engineers (CRE) Lines of Communication for the Eastern Mobile Force.

Rice arrived back in England, to disembark in August 1902, having served in the campaign with distinction. Mentioned in dispatches four times, he was awarded the Queen's South Africa Medal with two clasps and the King's South Africa Medal with two clasps. He returned to the School of Military Engineering in Chatham to resume his service, and at age 43 in 1903 he married Mary Dalrymple Scannell, who was born in Bombay, India, daughter of Major Dancar RA and widow of Mr Henry Scannell, in Kensington, London.[382]

His design for the 'Rice Pattern' Blockhouse is commemorated at Brompton Barracks in Chatham, Kent on the Royal Engineers South African War Memorial Arch. A panel high on the arch depicts a relief of the construction of a corrugated-iron blockhouse. The memorial was unveiled by King Edward VII, the then Colonel-in-Chief of the Royal Engineers, on 25th July 1905.

The Anglo-Boer War Memorial Arch, Chatham, Kent.
Courtesy of The Medway Archives Centre.

As a brevet colonel in October 1908, Colonel Rice, on completion of five years service as a regimental lieutenant colonel, was placed on half-pay.[383] He served in Chatham until 1909 when he was appointed Chief Engineer of the Coast Defences in the Southern Command. In 1911 he then became Chief Engineer of the Aldershot Command until 1914. The 1911 Census shows that Colonel Rice, aged 52 years, was living in Devonport with his wife of seven years, Mary Rice and four domestic servants at No. 5 Nelson Villas.

At the start of the First World War he was attached to headquarters units of the British Expeditionary Force as a colonel, appointed as temporary Brigadier-General, and for the purposes of pay graded as Chief Engineer,[384] serving mainly in France.

The Royal Engineers South African War Memorial Arch, Chatham, Kent, showing the left-hand panel depicting two mobile blockhouses and the right hand panel depicting the Rice Pattern Blockhouse.

On 3rd June 1913 he was awarded the Companion of the Most Honourable Order of the Bath by the King on the occasion of His Majesty's Birthday. At that time he was Colonel (temporary Brigadier-General), serving as Chief Engineer, Aldershot Command.[385]

On 26th January 1915 Spring Rice was appointed under the Headquarters appointment as Chief Engineer.[386] In 1916 he was appointed to the post of Brigadier-General Royal Engineers in I Corps, as the Engineer-in-Chief of the entire British Expeditionary Force, succeeding his old Anglo-Boer War colleague George Fowke, and was now serving under Lieutenant-General Sir Douglas Haig. During this period he worked on a trial project to form the Royal Engineers tunnelling companies, organising them into a distinct branch in the Corps.

From 1916 to 1917 he was Engineer-in-Chief of the headquarters units, and then took over command of the 4th Garrison. He was transferred to the Forts Garrison Command in 1917, a posting he held for the remainder of the war.[387] On 4th June 1917 he was awarded the Most Distinguished Order of Saint Michael and Saint George KCMG for his service in the Great War.[388]

On 20th July 1917, decorations and medals awarded by the Allied Powers at various dates to the British Forces for distinguished services rendered during the course of the campaign included 'His Majesty the King[389] gave unrestricted permission in all cases to wear the Decorations and medals in question. Decorations conferred by His Majesty The King Of The Belgians. The Ordre de Leopold, Commandeur.' After the end of the war he was awarded the French Croix de Guerre[390] and was by then Major-General Sir Spring Robert Rice, KCMG, CB. Sir Spring Rice retired from the Army in 1919 after having been mentioned in dispatches a further six times. He died suddenly on Sunday 11th August 1929 in Hove, Sussex, aged 71 years.

CHAPTER 10

THE TRANSITION TO PEACE

No age lives entirely alone; every civilisation is formed not merely by its own achievements but by what it has inherited from the past. If these things are destroyed, we have lost a part of our past, and we shall be the poorer for it.

BRITISH MONUMENTS MAN RONALD BALFOUR, 1944[391]

The Treaty of Vereeniging was signed at Melrose House in Pretoria, minutes before Lord Kitchener's midnight deadline, on 31 May 1902. At Vereeniging – literally at the eleventh hour – General Christiaan de Wet's change of mind resulted in the Boer delegates accepting the British terms for the cessation of hostilities, by a vote of 54 to 6. The British had inherited a virtual wasteland and the task of reconstructing this shambles was given to a man whom the Boers had considered to be a ruthless imperialist who was originally responsible for the war – Alfred Lord Milner, then High Commissioner for South Africa and Governor of the Orange River Colony and the Transvaal. To assist him in this monumental effort, Milner enlisted the services of a group of young Edwardian fellow Imperialists, all recent university graduates, who proudly accepted the name 'Milner's Kindergarten'. Their priority was the repatriation and re-settlement of displaced inhabitants from the former Boer republics.[392]

In modern parlance this is known as the 'post-conflict reconstructions phase', and at the time it was key for the Empire to establish a strong and settled South Africa, but the task was huge. Thousands of white and black families had to be resettled from the concentration camps, with the farm locations and means to support them all destroyed. Thirty thousand Boer homesteads burnt down during Kitchener's scorched earth policy, stripped of livestock and crops, had to be rebuilt. In terms of the Treaty, the British government made £3 million available to the Boers to cover the post-war resettlement programme and another £3 million in interest-free loans. Landowners were provided with farm implements and seeds in an attempt to quickly rebuild their farms and a

THE TRANSITION TO PEACE

sustainable food supply. Furthermore, another £2 million was made available as compensation to British subjects, neutral foreigners and Africans. The total eventually reached £16.5 million – a drastically lesser amount than the total cost of the war, which exceeded £200 million.[393]

BLOCKHOUSES SURPLUS TO REQUIREMENT

With over 9,000 blockhouses, forts, redoubts and earthworks erected by the British forces throughout South Africa, the question always arises – What happened to them? Part of the answer can be found in a notice that appeared in the *Northern Post* of 22 August 1902, only three months after the war ended, selling numerous items of military defence stores including two complete corrugated iron blockhouses. This heralded the wholesale dismantling and disposal of the blockhouse system: the backbone to Kitchener's grand strategy was now completely useless and disposable. It was to provide a key source of building and farming material in a resource-strapped economy desperate to rebuild and return to normal.

Advert for Important Sale of Fencing and Other Material in the Northern Post, 22 August 1902.

Typically these sales were conducted after blockhouses were dismantled and brought into towns or Royal Engineer Depots. All over the country auctions were conducted and local farmers bought corrugated iron, pickets, wire and complete blockhouses. The advertisement in the Aliwal North local paper lists a great deal of military materiel that would be ideal for farm repair and fencing in of cattle. At another sale a Mr Petrus Naude, from Brack River in the Victoria West District, bought a number of pieces of wood together with Nos 33 and 34 Blockhouses for £15 3s 6d.

Received from Mr P Naude the sum of fifteen pounds three shillings and six pence for	
No 36 & 37 Block House	£ 3. 10 0
@	£ 3. 10 0
And 29 pieces wood at 1/6	£ 2. 3 6
And 46 " " at 3/6	£ 6. 0 0
Signed as paid dated 6/12/02	£15. 3 6

Building material was very scarce after the war and every item old and new was being recycled into rebuilding farms and lives. Over 7,000 of these Rice pattern blockhouses were dismantled and absorbed into the devastated communities; it is hardly surprising that few remain intact today. There are many houses from this period all over South Africa still standing, and no doubt much of the old corrugated iron sheeting still survives. In your travels keep en eye open for the odd loophole that may still exist in in someone's house or shack out on a remote farmstead. Not all dismantling was carried out quickly and even a year later sales were still being carried out as described in the Krugersdorp *Standard* of 9 May 1903:[394]

For Sale:

This list contains the names of 22 districts and 920 blockhouses in the Transvaal. Tenders should be made so much per blockhouse either for the whole number offered, or for the number in any district. Purchasers will have right of access for the purpose of removal under Ordinance No. 39 of November 14th, 1902. The largest groups of blockhouses were 285 at Ermelo District and 158 at Vryburg District. Pretoria had 29.

Signed by T. Cadell
Assistant Secretary, For Secretary for Repatriation

Other structures were sold complete and transported intact, and survive today as relics of the past. At Hillston Farm in the Eastern Cape, a Burnaby Pattern blockhouse survives as a braai area for a farm Bed & Breakfast. After the war the farm's owners had also bought an old British Army barracks from a sale in Middelburg and reassembled it for use on the farm; today it acts as a sheep-shearing shed for the farm's flock of some 800 Merino sheep.

There were many visitors to South Africa after the war ended, tourists, those visiting relatives' graves and some to settle no doubt after falling in love with someone or just falling in love with Africa. One such traveller recounts:

We all went to have tea with them, and very kind and hospitable they were. We had tea at the dining-room table used by the Boers as their operating table! The walls are full of shell and bullet holes. The magistrate's daughter gave me some photos, taken by herself, of their house on the banks of the Vaal. A very interesting little blockhouse I found at Christiana had been turned by a cheery old Boer into a snug little home while waiting repatriation aid. He has ingeniously spread a tent over its top to make it rain-proof, and lives in it with his Frau, cultivating a pretty little garden around it, and drying the biltong, which you see hanging in strips to the tin sides of his blockhouse home. The repatriation here is conducted on an extensive scale, and the stores are so vast that the superintendent has his work cut out to protect them from theft. [Black] policemen and repatriation employees patrol them all night, and no one is allowed to approach the stores; which are in a barbed-wire enclosure on the open veldt.[395]

The repatriation of portable war stores was not the only activity; clearly there were the old stone blockhouses now also surplus to requirement, and here the disposal seemed less organised and advertised. Many would no doubt have been demolished for building material, especially those close to destroyed farms or on farm land. Gradually over the years the vast majority of those in what is now the sprawling metropolis of Gauteng have disappeared – originally used to protect the city, and now demolished for houses and factories generated by the vast mineral wealth of South Africa spread out over the open highveld.

Many blockhouses were sketched as facsimiles, painted or photographed for postcards sent home years after the war ended. A selection is shown here from the author's collection.

A selection of postcards from after the Anglo-Boer war depicting blockhouses, from top left clockwise: a Burnaby Pattern Blockhouse with the South Wales Borderers; the Orange Grove Octagonal blockhouse in Johannesburg; 'With Every Good Wish', a rectangular blockhouse in Ladysmith; and finally a sketch of a blockhouse by night and day.

Occasional modern day traces remain, such as Fort Street in a small Karoo town or the long-gone Kempton Park (Zuurfontein) blockhouse in Johannesburg, built to protect the station from marauding Boer commandos. Today its presence only exists as a name on a new building and the street name, Blockhouse Street. In more remote and rural areas, African families just moved in after the troops left, as people were in need of shelter from the rain or the sun. The postcard here shows one such family at some point after the war living in a rather dilapidated ad hoc blockhouse, most likely in the Cape region. In some areas there is evidence of more modern military occupation with recent regimental names and initials being left behind by occupying troops, such as those in the stone blockhouses in the Harrismith area.

A black family living in a post-war blockhouse.

HISTORICAL PRESERVATION

Then there was the matter of historical preservation, largely after the Union of South Africa was formed on 31 May 1910. Initially the desire to preserve prehistoric rock paintings and engravings encouraged foreign inbound expeditions to collect samples and remove them for museums abroad. In 1911 Parliament passed the first law to protect these ancient historical sites, the 'Bushmen Relic Protection Act'. By 1923 the limitation of this statute and the narrow focus of its protection had been realised and the 'National and Historical Monuments Act' was passed. Under the auspices of this act the Historical Monuments Commission was established. It was not empowered to recommend to the government that any building, site or relic should be proclaimed, but rather depended on the goodwill and funds of owners for the preservation and maintenance of monuments, which was far from ideal.

The Commission did carry out valuable work regardless and compiled lists of all old buildings, military fortifications, and other places noteworthy of protection and preservation. It was at this time that some restoration work was completed and the bronze badges started to appear on the designated structures. By 1934 both separate acts were replaced with the 'Natural and Historical Monuments, Relics and Antiques Act', and the Commission was replaced by a National Monuments Council, to which it handed over some 300 sites marked by signs and plaques and fenced off for limited protection.

Many other organisations have also been involved in maintaining the country's national heritage, including provincial administrations, municipalities, local historical societies, landowners and charities. The National Monuments Council was replaced by the South African Heritage Resources Agency (SAHRA) and the National Heritage Council in the SAHRA Act of 1999, both of which are still in place at the time of writing.

The key change is now that 'all previous national monuments became provincial heritage resources' with the 1999 Act and are managed by the Provincial Heritage Resources Authority, and that SAHRA is only managing 'national' heritage sites.[396] It now appears that the historic built environment of South Africa declared over a period of years since 1911 has been left to the provinces or individual owners to maintain, while the national body, SAHRA, has the onerous duty of caring for everything declared since 1999, including major monuments such as the Castle of Good Hope.

SAVING THE BLOCKHOUSE HERITAGE

As one might expect, the corrugated iron Rice Pattern blockhouses have all but disappeared and only seven exist today, representing less than one per cent of the 9,000 built for the war. Despite extensive further research no more could be added to this tally in twelve years of searching; maybe there are more still out there on the remote farms of South Africa? Those found are safe at least for now, two being national monuments and the other three on private land and part of the tourism sector. A further six replicas add significantly to this tally and most were passionately, if not always accurately, reconstructed as part of attractions built for the centenary celebrations of 1999–2002.

The stone-built blockhouses fared much better and approximately 66 out of the 441 Royal Corps of Engineers recorded blockhouses have survived in some form today, representing 15% of the wartime build. Of these 28 are listed as national monuments. During the author's visit to each monument site a survey was undertaken to establish the condition and a grading given. These are shown in the full blockhouse database in *The Field Guide to the Anglo-Boer War Blockhouses* companion guide to this book, which lists over 90 original blockhouses and replica sites. It is assessed that the preservation of these monuments owes more to luck and site remoteness than preservation effort. While the Acts of Parliament actually stipulate these structures must be preserved, the overall achievement is fair considering the lack of government funds allocated and their lack of importance in the new Republic of South Africa, with different priorities for its heritage. That aside, sites such as Uniondale and Warrenton are now complete ruins when they ought to be well-preserved monuments.

THE FUTURE OF BLOCKHOUSES AS MONUMENTS

The sad fact is that there is not enough money in the heritage budget to go around. The Anglo-Boer War did benefit from centenary status from 1999 to 2002 and much was done during these celebrations to highlight the era and preserve this thin three-year slice of South African military and social history. Since this time the plight of the nation and its continued journey to overcome Apartheid and now to overcome education, jobs and poverty issues means that preserving old blockhouses for the sake of heritage comes well down on the priority list.

The author's view however is that these neglected resources, with careful upgrading and resourcing, could be a much more credible and profitable part of the already established tourist industry. Carefully restored stone blockhouses could become roadside cafes, more remote ones, overnight stops for mountain bikers or stops on battlefield tours. Perhaps a small section of a blockhouse line could be rebuilt with several blockhouses and the connecting trenches and wires built as a tourist attraction, close to an existing battlefield site, similar to that rebuilt on the Somme for First World War trench systems? This key heritage needs to be incorporated into industry or existing infrastructure and used to create opportunities and jobs wherever possible. With limited funds and some imagination this could be achieved, and the secured heritage used to improve lives and provide a lasting legacy to the history of the so-called 'Last of the Gentleman's Wars', but which in fact scared the local population immeasurably.

One of several blockhouses constructed in Warrenton to protect the railway bridge; top shows how the blockhouse used to look in c1995[397] and above, on a field trip in 2013, it was a ruin with anything useful stripped out.

Appendix A: Blockhouse Lines Constructed During the Anglo-Boer War

Type	From	To / via	To	Finished	Distance Miles	BH Qty	BH per Mile	Date	Distance	Quantity
TOWN	Pretoria Town			Jan-01	0	60	0.00	N/A	N/A	Ave Estimate
SAC	Petrusburg	Abraham's Kraal	Glen Siding	Jun-01	62	62	1.00	Map	Map Measured	Ave Estimate
SAC	Petrusburg	Jacobsdal	Riversford	Jun-01	90	90	1.00	Map	Map Measured	Ave Estimate
RAIL	Barberton	Kaapmuinden Ext.		Jun-01	34	60	0.56	Map	Map Measured	Ave Estimate
RAIL	Cape Town	Beaufort West		Jul-01	310	554	0.56	Map	Map Measured	Ave Estimate
RAIL	Beaufort West	De Aar	Crocodile Pools	Jul-01	529	945	0.56	Map	Map Measured	Ave Estimate
RAIL	Naauwpoort Jnc	Vereeniging	Pietersburg	Jul-01	624	1 115	0.56	Map	Map Measured	Ave Estimate
RAIL	Naauwpoort Jnc	Hanover Road	De Aar	Jul-01	74	132	0.56	Map	Map Measured	Ave Estimate
RAIL	Cape Town	Muizenberg	Simon's Town Extn	Jul-01	23	23	1.00	Rusi Map	Map Measured	Ave Estimate
RAIL	Kraaifontein Jnc	Malmsbury Ext		Jul-01	30	30	1.00	The Times	Map Measured	Ave Estimate
RAIL	Bellville Junc	Sir Lowry's Pass Extn		Jul-01	35	35	1.00	Map	Map Measured	Ave Estimate
RAIL	Klapmuts Stn	Stellenbosch	Eerste River	Jul-01	16	16	1.00	Map	Map Measured	Ave Estimate
RAIL	Worcester	Robertson	Swellendam Ext	Jul-01	65	65	1.00	The Times	Map Measured	Ave Estimate
RAIL	Naauwpoort Jnc	Alicedale Jnc		Jul-01	185	185	1.00	Map	Map Measured	Ave Estimate
RAIL	Springfontein	Bethulie	Stormberg Jnc	Jul-01	98	175	0.56	Map	Map Measured	Ave Estimate
RAIL	Smatdeel	Winburg		Jul-01	20	36	0.56	Map	Map Measured	Ave Estimate
RAIL	Rosmead	Cradock		Jul-01	50	33	1.50	Estimate	Map Measured	Ave Estimate

Type	From	To / via	To	Finished	Miles	Qty	Mile	Date	Distance	Quantity
RAIL	Stormberg Jnc	Rosmead Jnc		Jul-01	83	147	0.56	Map	Map Measured	Ave Estimate
RAIL	Stormberg Jnc	Queenstown		Jul-01	66	117	0.56	Map	Map Measured	Ave Estimate
RAIL	Sterkstroom Jnc	Indwe Ext		Jul-01	55	98	0.56	Map	Map Measured	Ave Estimate
RAIL	Klerksdorp	Johannesburg	Springs	Jul-01	133	158	0.84	Map	Map Measured	REC BW 6
RAIL	Wolwehoek	Heilbron		Jul-01	29	51	0.56	Map	Map Measured	Ave Estimate
RAIL	Johannesburg	Standerton	Newcastle	Jul-01	183	327	0.56	Map	Map Measured	Ave Estimate
SAC	Villiersdorp	Heidelberg		Jul-01	38	38	1.00	Map	Map Measured	Ave Estimate
RAIL	Pretoria	Middelburg	Komatipoort	Jul-01	250	484	0.52	Bethell(11)	WO 78/5450	Ave Estimate
XC	Koopmans River	Modderfontein		Jul-01	55	55	1.00	Bethell Map	Map Measured	Ave Estimate
XC	Bloemfontein	Town		Aug-01	100	100	1.00	Map	Map Measured	Ave Estimate
XC	Naauwpoort	Pretoria		Aug-01	57	81	0.70	Map	Map Measured	Ave Estimate
RIVER	Bethulie	Aliwal North	Quaggafontein	Aug-01	80	80	1.00	Map	Bethell(29)	Ave Estimate
SAC	Modder River	Abraham's Kraal		Aug-01	65	65	1.00	Map	Map Measured	Ave Estimate
SAC	Heidelberg	Springs	Erste Fabriken	Aug-01	56	56	1.00	Map	Map Measured	Ave Estimate
TOWN	Johannesburg Town			Sep-01	0	50	0.00	REC BW 6	N/A	REC BW 6
SAC	Springs	Vlakplaats		Sep-01	23	23	1.00	Map	Map Measured	Ave Estimate
SAC	Greylingstad	Vlakplaats	Wilge River	Sep-01	63	63	1.00	Map	Map Measured	Ave Estimate
XC	Machavie	Lace Dia Mine		Oct-01	48	69	0.70	Map	Map Measured	Ave Estimate
XC	Kroonstad	Lace Dia Mine		Oct-01	17	24	0.70	Map	Map Measured	Ave Estimate
XC	Heilbron	Frankfort		Oct-01	34	20	0.60	Map	Map Measured	Ave Estimate
RIVER	De Lange Drift	Villiersdorp	Coal Mine Drift	Oct-01	186	186	1.00	Map	Map Measured	Ave Estimate

APPENDIX A

Type	From	To / via	To	Finished	Miles	Qty	Mile	Date	Distance	Quantity
XC	Wakkerstroom	Castle Neck	Derby (Piet Retief)	Oct-01	70	100	0.70	Map	Bethell(8)	Bethell(8)
SAC	Boshof	Vet River Bridge	Vet River Bridge	Nov-01	94	94	1.00	Map	Map Measured	Ave Estimate
XC	Dundee	Vryheid Extn		Nov-01	40	73	0.55	Map	Bethell(23)	Bethell(23)
SAC	Villiersdorp	Greylingstad		Nov-01	20	20	1.00	Map	Map Measured	Ave Estimate
SAC	Val Station	Brugspruit	Groot Oliphants	Nov-01	68	68	1.00	Map	Bethell	Ave Estimate
SAC	Pretoria (SW loop)			Nov-01	20	20	1.00	Map	Map Measured	Ave Estimate
XC	Potchefstroom	Scandinavia Drift	Leeuw Spruit	Nov-01	60	86	0.70	Map	Map Measured	Ave Estimate
RAIL	Bloemfontein	Sanna's Post		Dec-01	20	17	0.56	Map	Bethell(30(i))	Bethell(30(i))
RAIL	Sanna's Post	Thaba 'Nchu		Dec-01	20	25	0.80	Map	Bethell(30(ii))	Bethell (30(ii))
TOWN	Thaba 'Nchu	Town		Dec-01	0	10	0.00	N/A	N/A	Bethell(30)
XC	Thaba 'Nchu	Ladybrand (Maseru)		Dec-01	42	37	1.14	Map	Bethell (30(iii))	Bethell (30(iii))
XC	Chunespoort	Strydpoort		Dec-01	0	12	0.00	Map	N/A	REC BW 5
XC	Bethlehem	Harrismith		Dec-01	50	136	0.37	Map	Bethell(3)	Bethell(3)
XC	Bethlehem	Bethlehem Bridge		Dec-01	4	17	0.24	Map	Bethell(3)	Bethell(3)
XC	Elands River	Elands R Bridge		Dec-01	10	20	0.50	Map	Bethell(4)	Bethell(4)
RIVER	Bothaville	Klerksdorp		Dec-01	50	71	0.70	MAP	Map Measured	Ave Estimate
SAC	Villiersdorp Zand Bridge	Val Station Standerton Line		Dec-01	30	43	0.70	Map	Map	Ave Estimate

Type	From	To / via	To	Finished	Miles	Qty	Mile	Date	Distance	Quantity
XC	Wonderfontein	Carolina		Jan-02	25	55	0.45	Bethell(10)	Bethell(10)	Bethell(10)
XC	Victoria Road	Carnarvon		Jan-02	80	115	0.70	Map	Bethell(24)	Bethell(24)
XC	Lambert's Bay	Clanwilliam		Jan-02	34	53	0.64	Map	Bethell(28)	Betthell(28)
XC	Ventersdorp	Mooi River		Jan-02	19	22	0.86	Bethell(15)	Bethell(15)	REC BW6
XC	Ventersdorp	Tafelkop		Jan-02	27	62	0.44	Bethell(17)	Bethell(17)	Bethell(17)
XC	Kroonstad	Lindley		Jan-02	50	108	0.46	Bethell(1)	Bethell(1)	Bethell(1)
XC	Kaalfontein	Valsch R. Bridge		Jan-02	0	8	0.00	Bethell(1)	N/A	N/A
RAIL	Van Reenen's Pass	Ladysmith		Jan-02	20	47	0.43	Map	Map	REC BW 2
XC	Van Reenen's Pass	De Beer's Pass		Jan-02	20	31	0.65	Map	Bethell(5)	Bethell(5)
XC	De Beer's Pass	Cattle Pass		Jan-02	35	35	1.00	Map	Map Measured	Ave Estimate
XC	Bethlehem	Ficksburg (Brindisi)		Jan-02	40	47	0.85	Map	Bethell(7)	Bethell(7)
SAC	Hoopstad	Vet River Bridge		Jan-02	69	77	0.89	WO 108/131	Map Measured	Ave Estimate
XC	Machadodorp	Lydenburg Extn		Jan-02	45	55	0.82	BW7	Bethell(21)	Bethell(21)
XC	Standerton	Ermelo		Jan-02	56	137	0.41	Map	Bethell(9)	Bethell(9)
XC	Frankfort	Botha's Pass		Jan-02	90	187	0.48	Map	Bethell	Bethell
XC	Botha's Pass	Langkrans Nek (Botha's Pass)		Jan-02	20	14	0.70	Map	Map Measured	Ave Estimate
XC	Vaal River	Klerksdorp	Ventersdorp	Jan-02	50	59	0.85	Map	Bethell(16)	Bethell(16)
XC	Olifants Nek	Naauwpoort	Tafel Kop	Feb-02	50	71	0.70	Map	Bethell(18)	Bethell(13)
XC	Ermelo	Carolina		Feb-02	34	85	0.40	Bethell(9)	Bethell(9)	Bethell(9)
XC	Lichtenberg	Buffelsvlei		Feb-02	40	81	0.49	Bethell(19)	Bethell(19)	Bethell(19)

APPENDIX A

Type	From	To / via	To	Finished	Miles	Qty	Mile	Date	Distance	Quantity
XC	Lindley	Bethlehem		Feb-02	80	94	0.85	Map	Bethell(2)	Bethell(2)
TOWN	Ladysmith Town			Feb-02	0	22	0.00	REC BW 2	N/A	REC BW 2
RIVER	Lace Diamond Mine	Bothaville		Feb-02	40	57	0.70	Map	Map Measured	Ave Estimate
XC	Volksrust	Wakkerstroom		Feb-02	18	21	0.86	Map	Bethell(7)	Bethell(7)
XC	Bothaville	Cdo Drift (Valsch R) Extn		Mar-02	19	27	0.70	Map	Map Measured	Ave Estimate
XC	Ermelo	Amsterdam		Mar-02	45	32	0.70	Bethell(12)	Map Measured	Ave Estimate
XC	Mafeking	Lichtenburg		Mar-02	43	72	0.60	Map	Map Measured	Ave Estimate
RIVER	Commando Drift	Bothaville		Mar-02	20	33	0.60	Map	Map Measured	Ave Estimate
TOWN	Vryheid Town			Mar-02	0	24	0.00	REC BW 2	N/A	REC BW 6
TOWN	Dundee Town			Mar-02	0	6	0.00	REC BW 3	N/A	REC BW 7
TOWN	Pietersburg			Mar-02	0	27	0.00	REC BW 5	N/A	REC BW 5
XC	Calvinia	Clanwilliam		Apr-02	96	159	0.60	Bethell(27)	Map Measured	REC BW 5
XC	Carnarvon	Williston		Apr-02	81	131	0.62	Map	Map Measured	Original Map
XC	Harrismith	Oliver's Hoek		Apr-02	24	40	0.60	Map	Bethell(6)	Ave Estimate
XC	Olivier's Hoek	Allison's Kop		Apr-02	11	25	0.44	Map	Map Measured	REC BW 2
XC	Kimberley	Boshof		Apr-02	34	48	0.70	Map	Map Measured	Ave Estimate
RIVER	Bloemhof	Hoopstad		Apr-02	21	30	0.70	Map	Map Measured	Ave Estimate
XC	Maretzani	Polfontein		Apr-02	26	43	0.60	Map	Map Measured	Ave Estimate
XC	Williston	Calvinia		May-02	73	121	0.60	Map	Map Measured	Ave Estimate
			Blockhouses	TOTALS	5,897	8,931				

Abbreviations:

Bethell	*The Blockhouse System in The South African War*, pp282-85, Record of lines (No.)
BH	Blockhouse
Cdo	Commando
Dia	Diamond
Ext	Extension
Jnc	Junction
RAIL	Railway blockhouse line
REC BW	Royal Engineers Chatham (Library and Archives), Boer War Diaries (No.)
RIVER	River bank blockhouse line.
SAC	South African Constabulary posts
Stn	Station
WO	War Office
XC	Cross-country

Appendix B
Calculation of Cost of Blockhouse Materials & Labour

1. Cost of the Blockhouse System (Conservative)

a. The Basic Blockhouses (circa 8,000)

Type	Qty(1)	Ave Cost(2)	Total	
Masonry	441	£700	£308,700	
CGI	6,803	£38	£258,514	
Works	555	£30	£16,650	
	7,799		£583,864	£583,864

b. Blockhouse Barbed Wire

	Yards	GBP/Yard(3)	Total	
6-Strand Fence	1,166	£0.028	£32.65	
Blockhouses			7,799	
			£254,622	£254,622

c. Interconnecting Barbed Wire

	Miles	Cost/Mile	Total	
	3 700	£50.00	£185,000	£185,000

d. Other Based on 60% Use of total 7,799 Blockhouses

Material & Labour	Qty	Cost	Total	
Rifle Racks	4,679	£5.00	£23,397	
Alarms	4,679	£2.00	£9,359	
	Miles	Cost		
e. Telegraph Line	3,700	£5.00	£18,500	
			£51,256	£51,256
				£1,074,742

2. Cost of the Blockhouse System (Realistic)

a. The Basic Blockhouses

Type	Qty	Ave Cost (2)	Total	
Masonry	441	£900	£396,900	
CGI	8,124	£50	£406,200	
Works	555	£40	£22,200	
	9,120		£825,300	£825,300

b. Blockhouse Barbed Wire

	Yards	GBP/Yard (3)	Total	
6-Strand Fence	1,300	£0.028	£36.40	
Blockhouses			9,120	
			£331,968	£331,968

c. Interconnecting Barbed Wire

	Miles	Cost/Mile	Total	
Blockhouse lines	5,896	£50.00	£294,800	£294,800

d. Other: Based on 60% Use of Total 9,120 Blockhouses

Material & Labour	Qty	Cost	Total	
Rifle Racks	5,000	£5	£25,000	
Alarms	5,000	£2	£10,000	
	Miles	Cost		
Telegraph Line	5,896	£5	£29,480	
			£64,480	£64,480
				£1,516,548

Labour used

Troops	Estimate	Men per BH	Total BHs	Low Total	High Total
British & Colonial	Low (1)	7	8,000	56,000	
	High (4)	9	9,120		82,076
Africans	Low (1)	3	8,000	24,000	
	High (4)	4	9,120		36,478
		TOTAL		80,000	118,554

Notes:
(1) Taken from Royal Engineer Summary at end of war.
(2) From Bethell, pp288-289; an average was taken.
(3) Bethell, p289 (Cost at port, no labour cost).
(4) Author's estimation.

Appendix C
Rice and Elliot Wood Pattern Blockhouses – Materials used.

Rice Pattern (with Caisson)
Dimensions from Bethell Plate XV - XVI

Caisson		Imperial (ft)	Metric (m)	R x R (m)	Volume (m/cu)
Outer (OCr)	Radius	8 foot 1.5 inch	2.476	6.13	23.10
Inner (ICr)	Radius	6 foot 9 inches	2.057	4.23	15.94
Caisson height (Rh)	Height	3 foot 11 inches	1.2		
Pi		3.14			
Volume Calculation	V = Pi*R2*H				
Density (Gravel)		1,522 (kg/m/cu)			
Mass Calculation	mass = density x volume				

Caisson fill	Volume (m/cu)	7.16
Caison fill	Mass (kg)	10,892

Shield		Imperial (ft)	Metric (m)	R x R (m)	Volume (m/cu)
Outer (OSr)	Radius	8 foot 1.5 inch	2.476	6.13	13.28
Inner (ISr)	Radius	7 foot 6 inch	2.286	5.23	11.32
Shield height (Sh)	Height	2 foot 3 inch	0.69		
Pi		3.14			
Volume Calculation	V = Pi x (RxR) x H				
Density (Gravel)		1,522 (kg/m/cu)			
Mass Calculation	mass = density x volume				

Shield fill	Volume (m/cu)	1.96
Shield fill	Mass (kg)	2,984

Revetment		Imperial (ft)	Metric (m)	R x R (m)	Volume (m/cu)
Outer (ORr)	Radius	9 foot 1.5 inches	2.881	8.30	31.12
Inner (OCr)	Radius	8 foot 1.5 inches	2.476	6.13	22.98
Revetments height	Height	3 foot 11 inches	1.194		
Pi		3.14			
Volume Calculation	V = Pi x (RxR) x H				
Density (Rock)		2,700 (kg/m/cu)			
Density (Rock)		1,250 (kg/m/cu)			
Mass Calculation	mass = density x volume				

Revetment volume (m/cu)	8.13
Sloping wall (1/2 volume)	4.07
Total volume (m/cu)	12.20
Revetment mass Soil (kg)	15,251
Revetment mass Rock (kg)	32,943

Surrounding Trench (Plate VIII)		Imperial (ft)	Metric (m)	R x R (m)	Volume (m/cu)
Outer (OTr)	Radius	14 foot 6 inches	4.419	19.53	84.06
Inner (ITr)	Radius	10 Feet	3.048	9.29	39.99
Trench depth (Td)	Depth	4 foot 6 inches	1.371		
Pi		3.14			
Volume Calculation		V = Pi x (RxR) x H			
Density (Soil)		1,250 (kg/m/cu)	Trench volume (m/cu)		44.07
Density (Rock)		2,700 (kg/m/cu)	Out-fill Soil mass (kg)		55,088
Mass Calculation		mass = density x volume	Out-fill Rock mass (kg)		118,991

Surrounding Fence Wire		Imperial (yds)	Metric (m)	2 x R (m)	Circum. (m)
Single-strand Wire fence (6WFr)		25	22.86	45.72	143.56
Six Strands		6			861.36
Spider Web (6 Circles) Estimate					200.00
Pi		3.14	Total blockhouse wire (m)		1,061.36
Circumference Calculation		C = 2 x Pi x R			

Inter-connecting Wire

Blockhouse lines distance (Author's Calculation)		5,896	miles
6- strand fence total wire needed		35,376	miles (Minimum)
8-strand fence total wire needed		47,168	miles (Maximum?)

Interconnecting counter-mobility trenches		Imperial (yds)	Metric (m)		Volume (m/cu)
1 x Trench appox (3 ft x 3 ft)	Width/depth	1	2.743		
Distance apart (Approx 1/2 mile)	Length	880	804.67		
Volume Calculation	W x D x L				
Density (Soil)	1,250		2 x Trench volume (m/cu)		4,414.42
Mass Calculation	mass = density x volume		Out-fill Soil mass (kg)		5,664

APPENDIX C

Elliot Wood Blockhouse

Dimensions from Bethell Plate XIV

		Feet	Metres	Volume		
Ground floor	Floor height (h)	7	2.13	H x OW x IW		
	Outer wall (OW)	20	6.1		79	cu/m
	Inner wall (IW)	14	4.27		39	cu/m
			Walls volume		40	cu/m
			Walls mass		109,137	Kg
First floor	Floor height (h)	7.5	2.28	H x OW x IW		
	Outer wall (OW)	20	6.1		85	cu/m
	Inner wall (IW)	16	4.87		54	cu/m
			Walls volume		31	cu/m
			Walls mass		83,064	Kg
Top floor	Floor height (h)	3.6	1.06	H x OW x IW		
	Outer wall (OW)	20	6.1		39	cu/m
	Inner wall (IW)	17	5.18		28	cu/m
			Walls volume		11	cu/m
Density Rock	2,700		Walls mass		29,701	Kg
Density Concrete	2,400					

Stone to Concrete Comparison

	Volume	Stone (kg)	Concrete (Kg)
Ground floor walls	40	109,137	97,011
First floor walls	31	83,064	73,834
Top floor walls	11	29,701	26,401
Total Mass		221,902	197,246

GLOSSARY

Agterryer — (Afrikaans) An assistant who accompanies his employer on horseback, especially when hunting or on commando. *Agterryers* were either conscripted by the Boers or joined the commandos voluntarily, where they guarded spare ammunition and looked after the horses, cooking, collecting firewood and loading firearms. Not only were these 'auxiliaries' used in a labour capacity, but they were also used in fighting.

Arc of fire — The scope or number of degrees of the field of fire a projectile weapon has when firing through an aperture in the wall of fortification, such as a loophole, gun port, crenel or embrasure.

Bagged — Succeeded in catching or killing; total 'bag' on a game shoot is the total number of birds in the game bag.

Banquette — A continuous step or ledge at the base of a parapet, on which defenders stood to fire over the top of the wall.

Barbette, Barbet, Barquette — An earthen terrace or platform situated inside the parapet or a rampart, upon which cannon were mounted so that they could be fired over a wall rather than through a gun port. A battery in this situation is called a *battery en barbe* (or barbet).

Bartisan, Bartizan — A turret which projected at an angle from a tower, a parapet or near a gateway. Used as a watchtower or a defensive position by utilising flanking fire. Bartizans were also used as strengthening buttresses or simply as decoration on later castles or castellated residences.

Bastion — A work projecting from the curtain wall of a fortification which commanded the foreground and the outworks. Designed to provide flanking fire to adjacent curtains and bastion. Bastion has been used to refer to the flanking towers of a castle as well as the arrow-headed bastions of the Italian bastion trace.

Batman	A soldier or orderly assigned to a commissioned officer as a personal servant or valet. The term originates from the obsolete *bat*, meaning 'pack-horse'. Thus the batman was in charge of the officers 'bat-horse' which carried the pack saddle with the officer's kit during a campaign.
Batter, Battering	(1) A wall with a receding slope from the ground upwards, narrowing at the top, is said to be battered. See plinth. (2) To use a siege engine or artillery to batter or strike repeatedly against a fortifications wall or gate to make a breach.
Battered base, Battered plinth	The projection at the base of a wall which sloped outwards. Also known as battering.
Battlement	The upper part of a fortifications wall from which defenders defended their position. The battlement or parapet was usually provided with crenels and merlons; the crenels were the openings and the merlons were the solid uprights. This arrangement allowed the defenders to fire upon attackers through the crenels while obtaining some protection from the returned enemy fire behind the merlons.
Besiege	To surround, invest, or to lay siege to a place or fortification.
Bitterender	To the bitter end: often refers to the completion of a long and difficult task, and sticking it out to the very end, as the die-hard Boers did, becoming the *Bittereinders*. This term originates, ironically for the Boers, from British sailing slang, and relates to the anchor cable, which was attached on deck to a 'bitt', hence when the anchor cable was fully paid out to its maximum length it was said to be at 'the bitter end'. (Robson, M, *Not Enough Room to Swing a Cat, Naval Slang and its Everyday Usage* (Conway, 2008), p13.)
Blockhouse	Originally a fortification used for seaward defence provided with shot-deflecting battlements, hand-gun and rifle ports and a single embrasure for a long range cannon, first used during the 16th century.

Brattice, Brattish, Brettice	A small stone gallery built out from a castle's parapet or wall on corbels but lacking foot boards. Used to reduce the dead-ground at the base of the wall by allowing defenders to drop missiles through the hole onto the enemy below.
Breastwork	An earthwork thrown up to breast height which provided protection to defenders firing over the crest of the work while in a standing position.
Bunker	A bunker is a defensive military fortification designed to protect people or valued materials from falling bombs or other attacks. Bunkers are mostly below ground, compared to blockhouses which are mostly above ground.
Buttress	A buttress is a horizontal support for a building usually made of brick or stone, sloping out to a wide base.
Bywoner	(Afrikaans) A labourer or farmer working another person's land; a squatter or share cropper.
Caisson	In engineering, a boxlike structure used in construction work underwater or as a foundation. For blockhouses, a bullet-proof sandwich wall or box with filling to provide protection.
Cantonment	Derived from the French word *canton* meaning corner or district, and is used extensively in South Asia where the term also describes permanent military stations. In military parlance a cantonment is a permanent fort or barracks.
Casemate	A fortified position or chamber or an armored enclosure on a warship from which guns are fired through embrasures.
Chamfered corner	A walled right-angled corner, usually cut off to 45 degrees to form a secondary small wall from which to provide covering fire and to prevent danger from spalling.

Citadel	A fortress in or near a city which was used to control the city and its inhabitants; providing a strong defensive position, and once the outer defences had fallen it could be used as a final refuge.
Corbel	A stone bracket projecting from the wall used to support an overlapping parapet or a roof or floor beam, similar to a bracket.
Counter-guerrilla warfare	Operations and activities conducted by armed forces, paramilitary forces, or non-military agencies against guerrillas.
Countermure	(1) A strengthening wall built in front of another wall; (2) A wall which was built behind that of another as a reserve defence.
Crawlway	A door similar to that in an igloo where bending and crawling is required to gain entrance.
Cremaille	(1) A zig-zag line of a fortification; (2) A sawtooth pattern on the inside line of a parapet.
Crenel, Crenelle	The part of a parapet which is indented is called the crenel, alternating with the solid uprights called merlons, which allowed the defenders to fire at the enemy while gaining protection from the merlons against the returned fire.
Crenellation	A parapet consisting of merlons and crenels. The 'Licence to Crenellate' was a royal licence giving permission to holder to build a fortification or to fortify a present building.
Dead-ground	Any ground where the opponent cannot be seen or attacked with direct fire weapons, affording them a level of protection.
Defilade	To secure a work or part against enfilade fire.
Ditch	A wide trench excavated along the outer perimeter of a fortification, which was utilised to impede the approach of an enemy force towards the walls. The ditch was either filled with water or left dry.

Doctrine	Fundamental principles by which the military forces guide their actions in support of objectives. It is authoritative but requires judgement in application.
Drift	South African term, from the Afrikaans *drif*, for a river ford or crossing.
Earthwork	A fortification which was made chiefly of earth, either for temporary or permanent use, for either defensive or offensive purposes, constructed by excavating and embanking earth.
Embrasure	The opening in a crenellation or battlement between the two raised solid portions or merlons, sometimes called a crenel or crenelle. In domestic architecture this refers to the outward splay of a window or arrow slit on the inside.
Enfilade	Defensive weapons fired from the flank of a work and directed along or across another, in order to create cross-fire and hit the enemy from the side.
Esprit de corps	(French) Literally 'spirit of the body' in French. A feeling of pride and mutual loyalty shared by the members of a group.
Fieldwork	A temporary work constructed by an army in the field, used to cover an attack on a fortification, or as protection against another enemy army, especially a relieving force.
Firing loop, Firing slit	Apertures in the walls of a castle which were used by defenders to discharge arrows and bolts though at the enemy.
Fontein	(Afrikaans) Fountain or spring.
Fort	A work established for the defence of a land or maritime frontier, of an approach to a town, or of a pass or river.
Fortification	(1) The act or art of fortifying a military position by means of defensive and/or offensive works; (2) A work or structure, used as a military position; a fortified place or position.

Fortified house	A civilian dwelling or other property fortified for local protection or as a defensive protection. Many Boer farms were fortified in this manner.
Fortress	A strong permanent fortification which may include a town.
Fougasse	An explosive mine placed in a hole cut in the ground, facing towards the enemy approaches. The explosive charge when detonated showers the attacker in stone shrapnel.
Gabion	A wickerwork basket which was filled with earth and used to build field works, such as revetments and parapets, also used by sappers as cover from musket shot; as they advanced their trench the gabion was rolled before them. Also called a hurdle.
Gable	(1a) The generally triangular section of wall at the end of a pitched roof, occupying the space between the two slopes of the roof; (1b) The whole end wall of a building or wing having a pitched roof; (2) A triangular, usually ornamental architectural section, as one above an arched door or window.
Ganger's hut	The ganger is one who oversees other workers, a Foreman; the foreman's hut.
Guerrilla warfare	Military and paramilitary operations conducted in enemy-held or hostile territory by irregular, predominantly indigenous forces.
Hensopper	(Afrikaans) Afrikaners who surrendered or defected to the English typically did so with their 'hands up', and were derogatorily referred to as *hensoppers* by the other Boers.
Human factors	Often referred to as ergonomics; used to describe interactions between three interrelated aspects: individuals at work, the task at hand and the workplace environment.
Keep	Safest and strongest part of the castle, usually at the centre or on top of the motte in a Motte & Bailey design.

Khazi	(IsiZulu) Slang for toilet, possibly being derived from the Cockney word 'carsey'. It is also speculated that it could come from the African language word, Zulu or Swahili, 'M'khazi' that is used to refer to a latrine.
Koppie, kopje, kop	(Afrikaans) Small hill or knoll (literally), in practice almost any size of hill.
Kraal	(Afrikaans) Livestock enclosure, native or farm.
Krijsraad	(Afrikaans) Boer Council of War.
Laager	(Afrikaans) (1) An encampment formed by a circle of wagons; (2) An entrenched position or viewpoint that is defended against opponents.
Look-out tower	Small tower structure protruding above the normal roof line to give extended height for the purpose of better observation.
Loop, loophole	A narrow opening in the wall or merlon of a fortification, through which missiles were discharged at the enemy; positioned to command the approaches, as well as protecting the weak spots. Designed to provide the maximum amount of protection for the defenders as well as give a reasonable field of fire.
Machicolation	A series of openings provided by: building the parapet out on consoles projecting beyond the face of the wall; the space between the consoles providing the openings for the machicolations. Projectiles and liquids could be thrown onto the enemy at the base of the walls, thus reducing the dead-ground. The term is usually reserved for the stone constructions which in many countries superseded the wooden hoardings used for the same purpose about the middle of the 14th century.
Martello tower	Often known simply as Martellos, small defensive forts that were built across the British Empire during the 19th century, from the time of the Napoleonic Wars onwards.
Meeting engagement	A combat action that occurs when a moving force, incompletely deployed for battle, engages an enemy at an unexpected time and place.

Merlon	(1) The portion of a battlemented parapet that rises up from a wall (eg the solid part of a parapet between the crenels); (2) The part of a parapet between two embrasures.
Motte & Bailey	The motte is the defensive mound on which sits the keep; the bailey is the defensive perimeter in which the defending occupants reside.
Mutual support	That support which units render each other against an enemy, because of their assigned tasks, their position relative to each other and to the enemy, and their inherent capabilities.
Nascent	Emerging, just coming into existence, new.
Nek	(Afrikaans) Mountain pass.
Octogon, octagonal	An eight-sided building, shape or fortification.
Outwork	(1) The surrounding outer wall of a fortification; (2) Defences constructed beyond the line of the main works, designed to keep the enemy at a distance because of the effect of the enemy's projectile weapons (eg siege equipment or artillery); to break up the line of an assault; to cover the approaches to the other outworks; and when taken by an enemy force, leave them totally open to fire from the main works.
Parapet	The top of a wall of either a fortification or fieldwork, either plain or battlemented. Used to provide protection to the defenders behind the wall. See battlement, crenel, embrasure, merlon.
Pillbox	A small, low fortification that houses machine guns, anti-tank weapons, etc. A pillbox is usually made of concrete, steel, or filled sandbags. The originally jocular name arose from their perceived similarity to the cylindrical and hexagonal boxes in which medical pills were once sold.
Pitch	The angle of a roof relative to the horizontal, usually deemed low-pitched or high-pitched.
Poort	(Afrikaans) Break in a range of hills, gorge, defile.

Reconcentrado	(Spanish) One of the rural non-combatants, who has been reconcentrated, specifically in Cuba, the Philippines, etc during the revolution of 1895–1898 by military authorities in areas surrounding the fortified towns.
Redoubt, redout, reduit	(1) A small work placed beyond the glacis, but within musket shot of the covert way, made in various forms, known as a detached redoubt; (2) A small work built in a bastion or ravelin of a permanent fortification; (3) An outwork or fieldwork, square or polygonal in shape without bastion or other flanking defences, sited at a distance from the main fortification, used to guard a pass or to impede the approach of an enemy force.
Revetment	Sloped structures, formed to secure an area from artillery, bombing, or stored explosives.
Rondavel	(Afrikaans *rondawel*), a westernised version of the African-style hut, usually round or oval in shape, traditionally made with stones, mud and thatch.
Sangar	A small temporary fortified position originally made up of stone, now built of sandbags and similar materials.
Sapper	*Sapeur* (French), a pioneer or military engineer who performs a variety of military engineering duties such as digging trenches or saps in siege warfare.
Scorched earth	Deliberate destruction of resources in order to deny their use to the enemy.
Shell scrape	A type of military earthwork long and deep enough to lie in the prone position in, designed to shield a single soldier from artillery, mortar and direct small arms fire, it is not intended to be used for fighting from.
Spruit	(Afrikaans) Stream or brook.
Spur buttress	A sloping support for a wall or tower.
Stockade	(1) A defensive barrier made of strong posts or timbers driven upright side by side into the ground; (2a) A similar fenced or enclosed area, especially one used for protection; (2b) A jail on a military base.

Stroom	(Afrikaans) Stream.
Turret	A small tower or bartizan, which was often placed at the angles of a castle, to increase the flanking ability, some only serving as corner buttresses.
Uitskud	(Afrikaans) 'Shaking out' – strip another soldier of his equipment.
Vault	An arched work of masonry, forming a roof over the casemates, galleries or other spaces.
Veld (Veldt)	(Afrikaans) Plain or open country.
Walled city	An entire town or city ringed by a strong wall as defensive perimeter.

GLOSSARY REFERENCES

Handwoordeboek van die Afrikaanse Taal.
Hellis, J (1 June 2001), 'Why Pillbox?', Pillbox Study Group. Accessed 10 September 2009.
Joint Warfare Publication 0-01. *UK Glossary of Joint and Multinational Terms and Conditions.*
Mitchell, CF, *Building Construction. Part 1. First Stage or Elementary Course,* 2nd ed, revised (London: Batsford, 1889), p48.
Rankine, WJM. *A Manual of Civil Engineering* (Griffin, Bohn, and Co, 1862), p385.
The Free Dictionary, http://www.thefreedictionary.com/
The World Health Organisation.
Urban Dictionary, http://www.urbandictionary.com/
Wikipedia.
Wyley, SF, *A Dictionary of Military Architecture Fortification and Fieldworks from the Iron Age to the Eighteenth Century,* http://www.angelfire.com/wy/svenskildbiter/madict.html#Acknowledgments/.

BIBLIOGRAPHY

Aitkin, DW, 'The British Defence of the Pretoria–Delagoa Bay Railway', *SAMHS Journal,* Volume 11, Number 3/4 (October 1999).

Aitkin, D, 'Oliver "Jack" Hindon, Boer Hero and Train Wrecker', *Military History Journal*, Volume 15, Number1 (June 2001).

Amery, LS, *The Times History of The War in South Africa 1899–1902,* 7 Volumes (London: Sampson Low, Marston and Company, Ltd, 1907).

Andrews, T and Ploeger, J, *Street and Place Names of Old Pretoria* (Pretoria: JL van Schaik, 1989).

Andrews, TE, *'No Shots Fired': Pretorians Participation in the Second Anglo-Boer War 1899–1902* (Tom E Andrews, 1999).

Atwood, R, *Roberts & Kitchener 1900-1902* (Pen & Sword, 2011).

Atwood, R, *The Life of Field Marshall Lord Roberts* (Bloomsbury, 2014).

Backhouse, JB, Lt Col, *With the Buffs in South Africa*, 1903 (Ray Westlake Military Books facsimile, 1989).

Baker, A, *Battle Honours of the British and Commonwealth Armies* (Ian Allan Ltd, 1986).

Barker, B, 'Folly and Foolishness: The Rise and Ruination of Blockhouses', *SA Country Life,* No 41 (November 1999).

Barthorp, M, *Queen's Commanders* (Osprey History, 2000).

Belfield, E, *The Boer War* (Concise Campaigns Series, 1920) (Trinity Press, 1995).

Bethell, EH, *The Blockhouse System in the South African War*, Paper XII (Professional Papers of the Corps of Royal Engineers, Volume XXX, 1904).

BlueBooks, HMSO: Cd 453, Cd 454, Cd 457, Cd 819, Cd 853, Cd 893, Cd 902, Cd 934, Cd 1789, Cd 1790, Cd 1791, Cd 1792.

Brandt, J, *The Petticoat Commando Or Boer Women In Secret Service* (Mills & Boon Limited, 1913; Project Gutenberg E-book).

Bryant, John, *26.2 – The Incredible True Story of the Three Men Who Shaped The London Marathon* (Metro Publishing, 2013).

Byrnes, JS, *Unexploded Ordnance Detection and Mitigation* (Springer Science & Business Media, 2009).

Clayton, A, *Martin-Leake: Double VC* (Pen & Sword, 1995).

Coldstreamer, *Ballads of The Boer War, Selected from the Haversack of Sergeant J Smith* (Grant Richards, 1902).

Constantine, RJ (ed), *New perspectives on the Anglo-Boer War 1899–1902 / Nuwe perspektiewe op die Anglo-Boereoorlog 1899–1902* (Bloemfontein: War Museum, 2013).

Constantine, RJ, 'The Guerrilla War in the Cape Colony during the South African War of 1899-1902. A Case Study of the Republican and Rebel Commando Movement' (University of Cape Town thesis, 1996).

Corvi, SJ and Beckett, IFW, *Victoria's Generals* (Pen & Sword, 2009).

Croll, M, *The History of Landmines* (Pen & Sword, 1998).

Davey, AM, *Town Guards of the Cape Peninsula 1901–1902* (Castle Military Museum, 1999).

De Swardt, B, *963 Days at the Junction, A documented history of Springfontein [OFS] during the Anglo-Boer War 1899–1902* (Fontein Books, 2010).

De Wet, CR, *Three Years' War* (Charles Scribner's Sons, 1902).

Diespecker, D, 'The Naming of Steinaecker's Horse', *Military History Journal*, Volume 10, Number 3 (June 1996).

Dodd, F, *Generals of the British Army* (Teeling Press, 2008).

Downham, J, Lieutenant Colonel, *Red Roses on the Veld. Lancashire Regiments in the Boer War 1899–1902* (Carnegie Publishing Ltd, 2000).

Doyle, AC, *The Great Boer War* (1902), (Project Gutenberg, 2009).

Duxbury, GR, *David And Goliath, The First War of Independence 1880–1881* (SA National Museum of Military History, 1981).

Farwell, B, *The Great Boer War* (Pen & Sword Military Books), (paperback reprint of 1976 original, 2009).

Field, R, *Frontier Forts of the American Frontier 1820–91: Central and Northern Plains* (Osprey Publishing, 2005).

Frederick, Major General Sir Maurice, KCB with a staff of Officers, *History of the War in South Africa 1899–1902,* Volume 1-4 (London; Hurst & Blackett Limited, 1906).

Frescura, F, *A Field Guide to the Architecture of Southern Africa* (In Press 2001).

Fuller, JFC, Major General, *The Last of the Gentleman's Wars, A Subaltern's Journal of the War in South Africa 1899–1902* (London: Faber & Faber, 1937).

Gillings, Ken, *The Aftermath of the Anglo-Boer War* (The South African Military History Society Lecture, 2008).

Girouard, EFC, *History of the Railway During the War in South Africa, 1899–1902* (HMSO, 1903).

Gooch, J, *The Boer War: Direction, Experience and Image* (MPG Books Ltd, 2000).

Grant, MH, *History of the War in South Africa, 1899–1902*, Volume 4 (London: Hurst & Blackett, 1910).

Hattingh, Johan and Wessels, André, *Britse Fortifikasie in de Anglo-Boereoorlog 1899–1902* (Bloemfontein: War Museum, 1999).

Hattingh, Johan and Wessels, André, 'Life in the British Blockhouses during the Anglo-Boer War 1899–1902', *The South African Journal of Cultural History*, Volume 13, Issue 2 (November 1999), p39-55.

Holden, Lieutenant Colonel RM, 'The Blockhouse System in South Africa', *Royal United Services Institution Journal*, Volume 46, No 290 (1902), pp479-489.

Jackson, M. *The Record of a Regiment of the Line: A Regimental History of the 1st Battalion Devonshire Regiment During the Boer War 1899–1902* (Hutchinson & Co, 1908, Project Gutenberg E-book).

Judd, D, *Someone Has Blundered: Calamities of the British Army in the Victorian Age* (Phoenix, 2007).

Kearsey, AHC, *War Record of the York & Lancaster Regiment 1900–1900* (George Bell & Sons, 1903).

Knight, I, *Boer Commando 1876–1902* (Osprey, 2004).

Laws, D, *Who Killed Kitchener? The Life and Death of Britain's Most Famous War Minister* (Biteback Publishing, 2019).

Lee, E, *To The Bitter End, A Photographic History of the Boer War 1899–1902* (Protea Book House, 2002).

Lowther, HC, Lieutenant Colonel CMG, MVO, DSO, *Scots Guards from Pillar to Post* (London: Edward Arnold, 1911).

Macdonald, Wade et al, *HMS HAMPSHIRE 100 Survey 2016, Survey Report* (2020).

McLeod, 'The Psychological Impact of Guerrilla Warfare on the Boer Forces during the Anglo-Boer War', University of Pretoria thesis (2004).

Meyer, JH, in collaboration with EP du Plessis, *Kommando-Jare, 'n Oud-stryder se persoonlike relaas van die Tweede Vryheidsoorlog* (Cape Town: Human & Rousseau, 1971), 344 pp.

Miller, Margaret, & Russell, Helen (eds), *A Captain of the Gordons: Service Experiences, 1900–1909* (London: Sampson Low, Marston, n.d.).

Military Times, 'Boers, Blockhouses and Barbed Wire, Kitchener's Counter Insurgency', *Military Times*, Issue 14 (November 2011).

Milne, R, *Anecdotes of the Anglo-Boer War, Tales from the 'Last of the Gentlemen's Wars'* (Helion, 2013).

Mills, G & Williams, D, *7 Battles that Shaped South Africa* (Cape Town: Tafelberg, 2006).

Moffett, Private EC, *With the Eighth Division: A Souvenir of The South African Campaign* (Knapp, Drewett & Sons Ltd, 1903).

Moore, AT, *Professional Paper of the Corps of Royal Engineers*, Royal Engineers Institute Occasional Papers, Volume XXX (Chatham, 1905).

Murray, N, *The Rocky Road to the Great War: The Evolution of Trench Warfare to 1914* (Potomac Books, 2013).

Nasson, B, *The War for South Africa: The Anglo-Boer War 1899–1902* (NB Publishers, 2011).

Nasson, B, *Uyadela Wen'osulapo, Black Participation in the Anglo-Boer War* (Ravan Press, 1999).

Netz, R, *Barbed Wire: An Ecology of Modernity* (Wesleyan University Press, 2004).

Oberholster, JJ, *The Historical Monuments of South Africa* (Rembrandt van Rijn Foundation for Culture, 1972).

Otto, JC, *Die Konsentrasiekampe* (Cape Town, 1954).

Pakenham, T, *The Boer War* (Jonathan Ball Publishers, 1979).

Pampilis, J, *Foundations of the New South Africa* (Cape Town: Maskew Miller Longman, 1991).

Peters, WH, 'The Architecture of the Blockhouses of the Anglo-Boer South African War, 1899–1902', Part 1: 'Standard Pattern', *Journal of the Institute of South African Architects* (May/June 2003), pp46-53.

Peters, WH, 'The Architecture of the Blockhouses of the Anglo-Boer South African War 1899–1902', Part 2: 'Rice-Pattern'. *Journal of the Institute of South African Architects* (July/Aug 2003), pp44-53.

Peters, WH, 'The architecture of the blockhouses of the Anglo-Boer South African War 1899-1902. Part 3: Conclusions. Blockhouse or blockhead?' *Journal of the Institute of South African Architects* (Sept/Oct 2003), pp40-43.

Peters, W, Rice-type blockhouses at Mooi River. *Journal of the KwaZulu-Natal Institute for Architecture,* Issue 3/2001, pp16-17.

Preller, GS, *Kaptein Hindon, Oorlogsaventure van 'n Baas Verkenner* (Pretoria: JL Van Schaik, 1916).

Pretorius, F, *Scorched Earth* (Cape Town: Tafelberg, NB Publishers, 2017).

Rankin, Reginald, *A Subaltern's Letters to his Wife* (London: Longmans Green & Co, 1901).

Reitz, D, *Commando* (London: Faber & Faber, 1930).

Robson, M, *Not Enough Room to Swing a Cat, Naval Slang and its Everyday Usage* (Conway, 2008).

Royal Commission on the War in South Africa, *Minutes of Evidence Taken Before the Royal Commission on the War in South Africa*, Volumes I, II and III (London: His Majesty's Stationery Office, 1903).

Royal Engineers Institute, *Detailed History of the Railways of the South African War 1899–1902* (Chatham: Macay & Co Ltd, 1904).

Schoeman, C, *The Historical Karoo: Traces of the Past in South Africa's Arid Interior* (Cape Town: Zebra Press, 2013).

Scholtz, GD, *Die Tweede Wryheidsoorlog 1899–1902, The Anglo-Boer War 1899–1902* (Protea Book House, 2000).

Shearing, T and D, *Commandant Johannes Lötter & His Rebels* (self-published, 10 June 1998). http://www.sawar.co.za/

Shearing, T and D, *General Jan Smuts & His Long Ride* (self-published, 21 November 2000). http://www.sawar.co.za/

Shearing, T and D, *Gideon Scheepers & the Search for His Grave* (self-published, 21 September 1999). http://www.sawar.co.za/

Singer, B, *Churchill Style: The Art of Being Winston Churchill* (Harry N Abrahams, 2012).

Spies, SB, *Methods of Barbarism, Roberts and Kitchener and Civilians in the Boer Republics: January 1900–May 1902* (Jonathan Ball Publishers, 1977, 2001).

St. Leger, S, Captain. *Mounted Infantry at War* (Galago, 1903).

Steenkamp. Wilem, *The Black Beret*, (Heilbron, 2016)

Steevens, GW, *With Kitchener to Khartum* (Edinburgh: Blackwood, 1898).

Stirling, J, *Our Regiments in South Africa 1899–1902: Their Record Based on Despatches* (William Blackwood & Sons, 1903).

Sun Tzu, *The Art of War* (Lionel Giles, 1910).

Transvaal Critic (13 March 1903), 'Who Invented the Block-House System?' *Pretoriana, Magazine of the Old Pretoria Society*, No 73 (1977), pp14-23.

Todd, P and Forham, D, *Private Tucker's Diary: The Transvaal War of 1899–1902 with the Natal Field Force* (Elm Tree Books, 1980).

Tomlinson, Richard, 'Anglo-Boer War Town Guard Forts in the Eastern Cape 1901–1902, *SA Military History Journal,* Volume 10, Number 2 (December 1995).

Tomlinson, Richard, 'Britain's Last Castles, Masonry Blockhouses of the South African War 1899-1902', *SAMHS Review,* Volume 10, Number 6 (December 1997).

Trew, Peter, *The Boer War Generals (eBook Partnership 2013).*

Valiunas, Algis, *Churchill's Military Histories: A Rhetorical Study* (Rowman & Littlefield Publishers, 2002).

Van der Waag, I, *The Boer War Army, Nation and Empire: South Africa and The Boer Military System* (University of Stellenbosch, 1999).

Van Heyningen, E, 'A Tool for Modernisation? The Boer concentration camps of the South African War, 1900–1902', *South African Journal of Science*, Article Number 242 (8 June 2010).

Van Heyningen, E, *Concentration Camps of the Anglo-Boer War: A Social History* (Jacana, 2013).

Van Vollenhoven, AC and Meyer, A, "'n Oorsig van die Militêre Fortifikasies van Pretoria (1880–1902)', *Military History Journal*, Volume 9, Number 3 (June 1993).

Viljeon, Ben, General, *My Reminiscences of the Anglo-Boer War* (London: Hood, Douglass & Howard, 1902),

Von der Heyde, N, *Guide to the Sieges of South Africa* (Struik Travel & Heritage, 2017).

Von der Heyde, N, *Field Guide to the Battlefields of South Africa* (Struik Travel & Heritage, 2013).

Walker, RF, 'The Boer War Diaries of Lt Col FC Meyrick (Part II)', *Journal of the Society for Army Historical Research*, Volume 73, Number 295 (1995), pp155-180. JSTOR, www.jstor.org/stable/44224958. Accessed 12 May 2020.

Waller, S, 'A History of RE Operations in South Africa 1899-1902', Chapter XIV, Blockhouses pp1-36, original manuscript, unpublished and undated (Royal Engineers Chatham Library & Archives).

Walton, A, *New Zealand Redoubts, Stockades and Blockhouses, 1840–1848*, DOC Science Internal Series 122 (Wellington: Department of Conservation, 2003).

Warner, P, 'Army Life in the '90s', *Country Life* (1975).

Warrick, P, *Black People and the South African War 1899–1902* (Cambridge University Press, 2004).

Watson, M, Colonel, Sir Charles, W, *History of the Corps of Royal Engineers*, Volume III (Royal Engineers Institute, 1915).

Webster, R, *At the Fireside: True South African Stories*, Volume 1 (New Africa Books, 2001).

Weir, R, 'A Note on British Blockhouses in Hong Kong', *Surveying & Built Environment*, Volume 22, Issue 1 (December 2012).

Welch, JC, 'The Blockhouse Line', *Journal of the Society for Army Historical Research*, Volume 83, Number 334 (2005), pp93-109. JSTOR, www.jstor.org/stable/44231167. Accessed 12 May 2020.

Wessels, André, *Boer Guerrilla and British Counter Guerrilla Operation in South Africa, 1899 to 1902* (Department of History, University of the Free State, 2011).

Wessels, André, *The Anglo-Boer War 1899–1902: White Man's War, Black Man's War, Traumatic War* (Sun Press, 2011).

Westby-Nunn, Tony, *A Tourist Guide to the Anglo-Boer War* (Self Published, 2000).

Wilson, HW, *After Pretoria: The Guerrilla War,* Volumes 1 & 2 (The Amalgamated Press, 1902).

Wilson, HW, *With the Flag to Pretoria: A History of the Boer War of 1899–1900,* Volumes 1 & 2 (The Amalgamated Press, 1902).

Wilson, John, *C.B.: A Life of Sir Henry Campbell-Bannerman* (London: Constable, 1973).

Wood, E, Major General, *Life and Adventure in Peace and War* (Edward Arnold & Co, 1924).

Wood, P, 'Re-Discovery of Boksburg Blockhouse Foundations', Boksburg History Society (October 2010).

Zaloga, SJ, *Armored Trains* (Osprey, 2008).

DIGITAL MEDIA

The Royal Engineers Corps History CD-ROM, The Institute of Royal Engineers (InstRE).

DVD 'The Boer War 1899–1902', DVD & Memorabilia Collection, Goentertain TV.

WEBSITES

Canadian Military Heritage Project: South African War (Boer War): http://www.rootsweb.ancestry.com/~canmil/boer/index.html?cj=1&o_xid=0001091115&o_lid=0001091115

Castles and Fortifications of England and Wales: http://www.ecastles.co.uk/

Locations of all Anglo-Boer War Concentration Camp sites: http://www.eggsa.org/library/main.php?g2_itemId=43

Locations of Monuments on Google Earth: www.vuvuzela.com/googleearth/monuments general.kmz

Military Architecture: http://www.militaryarchitecture.com

Pill-Box Study Group (UK): http://www.pillbox-study-group.org.uk

South African Cannon Association: http://www.sa-cannon.com/index.html

South African Heritage Resources Agency (SAHRA): http://www.sahra.org.za/

SA Museums on-line: http://www.museumsonline.co.za

The Fortress Study Group: http://www.fsgfort.com/

The South African Built Environment: http://www.artefacts.co.za/index.htm

ENDNOTES

CHAPTER 1: A BRIEF HISTORY OF BLOCKHOUSES
[1] 'Martello tower', http://en.wikipedia.org/wiki/Martello_tower 26 February 2014.
[2] Brock, BB and Brock, BG, *Historical Simon's Town* (Cape Town, 1976), p162.
[3] The Heritage Portal, *A Short History of Simon's Town's Famous Martello Tower*, http://www.theheritageportal.co.za/article/short-history-simons-towns-famous-martello-tower. Accessed 17 August 2016.
[4] Field, R, *Frontier Forts of the American Frontier 1820–91: Central and Northern Plains* (Osprey Publishing, 2005).
[5] Ibid, p5.
[6] Mitchell, J, 'Side Trips: Fort Fairfield Block House Preserves Era of Conflict in Northern Maine', Aug 21, 2014. http://news.mpbn.net/post/side-trips-fort-fairfield-block-house-preserves-era-conflict-northern-maine. Accessed 2 March 2015.
[7] Blockhouse (Lockmaster's house) at Rideau Narrows lock. Photo courtesy of D Gordon & E Robertson, http://commons.wikimedia.org/wiki/File:Blockhouse_at_Rideau_Narrows.jpg 17. Accessed February 2015.
[8] The St Andrews Blockhouse, New Brunswick. Photo courtesy of John Stanton. http://www.fortwiki.com/File:St_Andrews_Blockhouse_-_23.jpg.
[9] Walton, A, *New Zealand redoubts, stockades and blockhouses, 1840–1848*, DOC Science Series 122, Department of Conservation, Wellington, 2003.
[10] Ibid, p7.
[11] Wallaceville Blockhouse, New Zealand. https://en.wikipedia.org/wiki/Upper_Hutt_Blockhouse#/media/File:Upper_Hutt_Blockhouse2.JPG.
[12] McCracken, H, 'Upper Hutt Blockhouse', *The Register*. New Zealand Historic Places Trust/Pouhere Taonga, 2001. Accessed 12 January 2013.
[13] Duke, J, Lieutenant-Colonel, *Recollections of the Kabul Campaign, 1879 & 1880* (Pickle Partners Publishing, 2017).
[14] Woodburn, CW, Brigadier, *The Bala Hissar of Kabul. Revealing a fortress-palace in Afghanistan*, The Institute of Royal Engineers Professional Papers (2009), No1.
[15] British Library, by Bengal Sappers and Miners, 1879, Photograph 197/(21).
[16] Ibid, original diagram by Ian Templeton.
[17] Bloss, JFE, 'The Story of Suakin (Concluded)', *Sudan Notes and Records*, Volume 20, No 2 (1937), pp247-280. *JSTOR*, www.jstor.org/stable/41716263. Accessed 22 July 2020.
[18] Durham University Library and Collections SAD.A89/7.
[19] Davis, RH, *Cuba in Wartime* (New York, RH Russell, 1987). http://www.online-literature.com/richard-davis/cuba/ 6 February 2012.
[20] Ibid.
[21] Ibid.
[22] http://www.latinamericanstudies.org/1868.htm. Accessed 20 June 2015.

CHAPTER 2: A SYNOPSIS OF THE ANGLO-BOER WAR
[23] South Africa History On-Line: Slavery is abolished at the Cape. http://www.sahistory.org.za/dated-event/slavery-abolished-cape. Accessed 14 January 2014.
[24] The Victorian era of British history was the period of Queen Victoria's reign from 20 June 1837 until her death, on 22 January 1901.

[25] Milne, *Anecdotes of the Anglo-Boer War*, p14.
[26] Gross, DM, *99 Tactics of Successful Tax Resistance Campaigns* (Picket Line Press, 2014), p94.
[27] Duxbury, *David And Goliath*, p14.
[28] Ibid.
[29] Austin, B, 'Wireless in the Boer War', *The Journal of the Royal Signals Institution*, Volume XXV No1 (Spring 2004), p25.
[30] *Royal Commission on the War in South Africa: Minutes of Evidence Taken Before the Royal Commission on the War in South Africa*, Volumes I, II and III. (London: His Majesty's Stationery Office, By Wyman And Sons, Ltd 1903), p94.
[31] Pakenham, *The Boer War*, p41.
[32] Heunis, MC, Oranje Vrijstaaat Artillerie Corps, presentation to SAMHS (10 August 2006).
[33] Wessels, *Boer Guerrilla and British Counter Guerrilla Operation*, p29: map showing initial Boer and British positions in 1899.
[34] Ibid, p37: map showing the British offensive to December 1899.
[35] Reitz, *On Commando*, chapter 3.
[36] Wessels, *Boer Guerrilla and British Counter Guerrilla Operation*, p49: map showing the British offensive from February to September 1900, including relief of the three siege towns.
[37] Von der Heyde, *Field Guide to the Battlefields of South Africa*, p127.
[38] Atwood, *Roberts & Kitchener*, p209.
[39] Corvi & Beckett, *Victoria's Generals, p206.*
[40] Wessels, *Boer Guerrilla and British Counter Guerrilla Operation*, p9.
[41] Atwood, *Roberts & Kitchener*, p210.
[42] Wessels, *Boer Guerrilla and British Counter Guerrilla Operation.*
[43] Author's photograph taken from a postcard of the day of the Coldstream Guards in Middelburg, Cape Colony.
[44] Van Heyningen, *Concentration Camps of the Anglo-Boer War,* piii.
[45] Rankin, *A Subaltern's Letters to His Wife,* p91.

CHAPTER 3: BRITISH STRATEGY AND TACTICS
[46] Netz, *Barbed Wire*, p65.
[47] Pampils, *Foundations of the New South Africa*, p35.
[48] Aitkin, 'The British Defence of the Pretoria–Delagoa Bay Railway'.
[49] Girouard, *History of the Railways during the War in South Africa*, pp39-40.
[50] Ibid, p41.
[51] Ibid, p46.
[52] Hattingh and Wessels, *Britse Fortifikasie in de Anglo-Boereoorlog*, p140.
[53] Mills, IP Lt Col, 'Railway Attack! An IED of the Anglo-Boer War, 1899-1902', *SAMHS*, Volume 15, No1 (June 2010).
[54] Preller, *Kaptein Hindon.*
[55] Aitken, D, 'Oliver "Jack" Hindon, Boer Hero and Train Wrecker', *Military History Journal*, Volume 12, No 1 (June 2001).
[56] SANDF Photography Archives (Pretoria) Photograph 831000756.
[57] King's College London, Liddell Hart Centre, Hamilton Papers: 2/3/5 - 1901 Nov 29.
[58] *Royal Engineers Journal* (1 December 1903), p265.
[59] *The History of The Boer War*, Sampson Low, Marston and Company Ltd, London (1902/6, Author's Collection).
[60] CRE's Report, Pretoria District, 14 May 1902, in the Boer War Diaries section, Royal Engineers Library. The term 'redoubt' was also used to designate these early blockhouses.

61 Durham University Library and Collections. SAD.628/115.
62 Bethell, *The Blockhouse System in The South African War*, Plate XVII.
63 Girouard, *History of the Railway During the War in South Africa*.
64 Trew, *Boer War Generals*.
65 Spies, *Methods of Barbarism* (1977), p107.
66 Otto, *Die Konsentrasiekampe*, p26.
67 Spies, *Methods of Barbarism* (1977), p112.
68 Spies, *Methods of Barbarism* (2001), pp117-8; Van Heyningen, *Concentration Camps of the Anglo-Boer War*, p58.
69 Kitchener, Army Circular 29, 21 December 1900. http://www.scielo.org.za/scielo.php?script=sci_arttext&pid=S0018-229X2019000200001.
70 Author's calculations.
71 Figures for white camps from Van Heyningen, *Concentration Camps of the Anglo-Boer War*, p115; black camps from Pretorius, *Scorched Earth*, p122.
72 Van Heyningen, *Concentration Camps of the Anglo-Boer War*, piix.
73 Pretorius, *Scorched Earth*, pp127 and 149.
74 Van Heyningen, *Concentration Camps of the Anglo-Boer War*, p76.
75 Farwell, *The Great Boer War*.
76 Worsfold, B, *Lord Milner's Work in South Africa* (Hazel, Watson, & Viney, 1906), p569.
77 Van Heyningen, *Concentration Camps of the Anglo-Boer War*, p79.
78 Pakenham, *The Boer War*, p548.
79 The National Archives, London, Kitchener Papers, PRO 30/57/22, 13 Dec 1901 letter from K to Broderick.
80 Waller, *A History of RE Operations in South Africa*, pp33-35.
81 Corvi & Beckett, *Victoria's Generals*, p212.
82 Miller & Russell (eds.), *A Captain in the Gordons: Service Experiences 1900-1909*, pp130f, Letter of 25 April 1902.
83 Farwell, *The Great Boer War*.
84 Scholtz, *Die Tweede Wryheidsoorlog*.
85 Smith, 'Langverwacht 23-24 February 1902', *SAMHAS Journal*, Volume 13, No 6.
86 De Wet, *Three Years' War*, Chapter 31.
87 Farwell, *The Great Boer War*.
88 Author's own postcard collection.
89 King's College London, Liddell Hart Centre, Hamilton Papers 2/3/16 8 Feb 1902 letter to R from PTA.
90 Atwood, *Roberts & Kitchener*, p254.
91 Military Times, *Boers Blockhouses and Barbed Wire, Kitchener's Counter Insurgency*, Issue 14 November 2011.
92 The National Archives London, Kitchener Papers: PRO30/57/21 dated 5 Nov 1901.
93 De Wet, *Three Years' War*, Chapter 31.
94 Farwell, *The Great Boer War*, p355.
95 Ibid, p356.
96 De Wet, *Three Years' War*, Chapter 31.
97 http://en.wiktionary.org/wiki/nascent. Accessed 14 May 2014.
98 http://en.wiktionary.org/wiki/doctrine. Accessed 14 May 2014.
99 http://en.wikipedia.org/wiki/On_the_fly. Accessed 14 May 2014.
100 http://en.wikipedia.org/wiki/Military_doctrine. Accessed 14 May 2014.

[101] Royal Engineers Institute, *History of the Railways of the South African War*, p248.

[102] Pritchard, Major-General HL, 'Memoir of Sir Percy Girouard', *Royal Engineers Journal* 63, June 1933.

[103] By Studio Wn. Notman & Son, Public Domain, https://commons.wikimedia.org/w/index.php?curid=37344025.

[104] Moore-Morris, Ralph, 'The Sudan Military Railway', *Soldiers of the Queen* 94, September 1898, pp8-11.

[105] Girouard, *History of the Railway During the War in South Africa,* p25.

[106] Atwood, *Roberts & Kitchener,* p163.

[107] Photograph source: *Detailed History of the Railways in the South African War 1899-1902*, Volume II Illustrations, Chatham Royal Engineers Institute (1904).

[108] Viljoen, Ben, *My Reminiscences of the Anglo-Boer War*, pp385-86.

[109] Royal Engineers Chatham Boer War Diary 9. Search Light Section Report for Quarter ending March 1902.

[110] Bethell, *The Blockhouse System in The South African War,* Plate XVII.

[111] Amery (ed), 7, pp24-25.

[112] De Wet, *Three Years' War*, p23.

[113] Peace Treaty of Vereeniging, South African History Online, 20 May 2018.

[114] Geof Ziezulewicz 'Fast adaptation of MRAPs has military catching up', *Stars and Stripes*, 8 June 2010.

[115] Nick Clegg MP, 'The Terms of Intervention', transcript of a speech to Chatham House, 23 June 2008.

CHAPTER 4: EVOLUTION OF THE BLOCKHOUSE SYSTEM IN SOUTH AFRICA

[116] Southland Times, Issue 15182, 12 March 1902, The National Library of New Zealand, p4.

[117] http://capetown.hotelguide.co.za/History_of_Table_Mountain-travel/looking-back-table-mountain.html 26 August 2014

[118] Rudman, JGA, Rear Admiral (Retd), 'Notes on Martello Towers', Newsletter No 440, South African Military History Society, Cape Town Branch (March 2016).

[119] Wood, *Life and Adventure in Peace and War.*

[120] https://www.karooimages.co.za/.

[121] Laband, J and Thompson, P, *A Field Guide to the War in Zululand 1879* (Pietermaritzburg, 1979), p20.

[122] 'Eastern frontier(s) of the Cape of Good Hope colony, ca 1820 –1850', http://commons.wikimedia.org/wiki/File:Eastern_Frontier,_Cape_of_Good_Hope,_ca_1835.png#mediaviewer/File:Eastern_Frontier,_Cape_of_Good_Hope,_ca_1835.png.

[123] Wyley, SF, *A Dictionary of Military Architecture Fortification and Fieldworks from the Iron Age to the Eighteenth Century* (1990). http://www.angelfire.com/wy/svenskildbiter/madict.html#Introduction. Accessed 26 August 2014.

[124] Laband, JPC, 'British Fieldworks of the Zulu Campaign of 1879, with Special Reference to Fort Eshowe', *Military History Journal,* Volume 6, No 1 (1983).

[125] Emms, M, 'East Fort – Monument to the Vandals, Strubenskop', *Pretoriana Magazine* (1979).

[126] 'James Murray (VC)', https://en.wikipedia.org/wiki/James_Murray_(VC). Accessed 4 August 2020.

[127] *SAHRA Gazette* No 3072, dated 23-04-1971.

[128] Tomlinson, *Britain's Last Castles.*

[129] *The Manual of Military Engineering*, Section 17, 'Blockhouses, Field Defences', Part 1

(1894), p107.

[130] Jones, J, Sergeant-Major RE, *The Iron Banded Gabion and its applicability to Military Field Purposes* (AT Fordham, 1860).

[131] Wyley, SF, *A Dictionary of Military Architecture Fortification and Fieldworks from the Iron Age to the Eighteenth Century.*

[132] Royal Engineers Chatham, Boer War Diary 9. Commander Royal Engineers' Report, Pretoria District, 14 May 1902. The term 'redoubt' was also used to designate these early blockhouses.

[133] *Royal Engineers Journal*, 1 December 1900, p246.

[134] Tomlinson, *Britain's Last Castles*.

[135] Ibid.

[136] Durham University Library and Collections SAD.628/109.

[137] Pole Evans, Illyd Buller, Copyright SANBI, licensed under Creative Commons Licence CC-BY.

[138] Plans based on author's visit, March 2012.

[139] SANDF Photography Archive No. 771001450.

[140] Tomlinson, *Britain's Last Castles*.

[141] Photograph from the Shell Series of 370mm by 315mm cards; The South African War 1899-1902 No37. The title reads: 'In desperation, the British divided the country into zones and tried to protect railway lines and bridges with barbed wire and blockhouses – but this was not successful.' The set was most likely issued to commemorate the 100-year anniversary of the war.

[142] Durham University Library and Collections SAD.628/129.

[143] Tomlinson, *Britain's Last Castles*.

[144] Andrews, T and Ploeger, J, *Street and Place Names of Old Pretoria* (Pretoria: JL van Schaik, 1989).

[145] Photography sources: (1) SANDF Photography Archive No 991001190, unknown source; (2) Liddell Hart Centre for Military Archives (King's College London Archives) Image 4458/381; (3) Photograph source unknown, a similar blockhouse still exists in Danielskuil and is a National Monument; (4) National Army Museum (London) 1972-08-72-2-73 Negative No 134356.

[146] *Royal Engineers Journal*, 1930, p348.

[147] Royal Engineers Chatham, Boer War Diary 8.

[148] Bethell, 'The Blockhouse System in The South African War', p278.

[149] Millicent Garrett Fawcett, https://commons.wikimedia.org/wiki/File:Millicent_Fawcett.jpg

[150] London School of Economics: The Women's Library, Papers of Millicent Garrett Fawcett, Item 7MGF/E/1/012 c.1901.

[151] *Royal Engineers Journal*, 1930, p348.

[152] Bethell, 'The Blockhouse System in The South African War', p279.

[153] National Army Museum, Chelsea, 2000-12-55-1 Negative No.133537.

[154] Royal Engineers Chatham, Boer War Diary 6 February 1901.

[155] Royal Engineers Chatham, Boer War Diary 8.

[156] Royal Engineers Chatham, Boer War Diary 2.

[157] REC Photograph Album No.1 annotated 'Rice B.H. near Lindley'. Canvas roof only supplied at first so as to save transport.

[158] The National Archives, Kew, WO108/347 CRE Natal Report of 6 June 1902 photograph annex.

[159] SANDF Photographs Nos. 761005277 and 761005276.

160 Farwell, *The Great Boer War*.
161 Royal Engineers Chatham, Boer War Diary 8, 23 Fd Company RE Notes 1899–1902 (Dated 31 May 1902 Middelberg Tvl), by Major Spring Rice RE, CRE L of C East, p25.
162 The National Archives London. WO/108/347.
163 *Royal Engineers Journal* (1 December 1903), p268.
164 Royal Engineers Chatham, Boer War Diary 1. Commander Royal Engineers' Army Report (13 May 1902), signed by Elliot Wood.
165 Royal Engineers Chatham, Boer War Diary 3.
166 Wilson, *After Pretoria*.
167 Lloyd, A, *The Zulu War* (History Book Club, 1973), pp46-47.
168 Walker, 'The Boer War Diaries of Lt Col. FC Meyrick'.
169 Bethell, 'The Blockhouse System in The South African War', p282.
170 *Royal Engineers Journal* (January 1902), p4.
171 Author's generic diagram based on historical accounts.
172 Kearsey, *War record of the York & Lancaster Regiment*, p188.
173 Royal Engineers Chatham, Boer War Diary 1.
174 Bethell, 'The Blockhouse System in The South African War', p289
175 Royal Engineers Chatham, Boer War Diary 1, Table of Blockhouses & post built up to 1902, at the end of the war.
176 Map, Transvaal, Spottiswoode & Co Ltd, London and Appendix A calculations: Machavie to Lace Diamond Mine 48 miles (69 Bhs); Kroonstadt to Lace Diamond Mine 24 Miles (17 Bhs); Potchefstroom to Leeuw Spruit 86 Miles (60 Bhs); Ventersdorp to Mooi River 19 miles (22 Bhs).
177 King's College London, Liddell Hart Centre, Hamilton Papers : 2/3/4 - 1901 Nov 26-1901 Pretoria 2 Dec 1901, three telegram transcripts.
178 National Army Museum Archive, NAM 1972-05-17-11.
179 http://www.fortadams.org/terminology.htm. Accessed 17 November 2014.
180 Wyley, SF, *A Dictionary of Military Architecture, Fortification and Fieldworks from the Iron Age to the Eighteenth Century*. http://www.angelfire.com/wy/svenskildbiter/madict.html#Acknowledgments.
181 Tomlinson, 'Anglo-Boer War Town Guard Forts In The Eastern Cape'.
182 Davey, *Town Guards of the Cape Peninsula*, p22.
183 Ibid, p29.
184 Ibid, p86.
185 National Archives, London, WO100/286.
186 http://angloboerwar.com/forum/2-introductions/1171-montagu-town-guard?limitstart=0 . Accessed 4 April 2013.

CHAPTER 5: BLOCKHOUSE SYSTEM COMPONENTS
187 Fuller, *The Last of the Gentleman's Wars*.
188 King's College London, Liddell Hart Centre, Hamilton Papers 2/3/17. 21 Feb 1902 letter to Lord Roberts.
189 Ibid.
190 Fuller, *The Last of the Gentleman's Wars*.
191 Original photograph, Author's collection, annotated 'Jim in the Canadian Mounted Rifles 1902'.
192 Warwick, P. *Black people and the South African War*, p23.
193 Atwood, *Roberts & Kitchener in South Africa*, p244.
194 Walker 'The Boer War Diaries of Lt Col FC Meyrick', pp155-180.

ENDNOTES

[195] Bethell, 'The Blockhouse System in the South African War', Plate VIII. Photograph courtesy of the War Museum, Bloemfontein.

[196] Ibid, Plate IX.

[197] Ibid, Plate II. Photograph courtesy of the War Museum, Bloemfontein.

[198] Liddell Hart Centre for Military Archives (King's College London Archives) Image: 4458/367.

[199] Kearsey, AHC, War Record of the York & Lancaster Regiment 1900–1900, p188.

[200] The National Archives, London WO 108/130 Blockhouse progress reports date: 1901 Dec –1902 Mar.

[201] Bethell, 'The Blockhouse System in The South African War' Plate VIII. Photograph courtesy of ABW Museum, Bloemfontein.

[202] Royal Engineers Chatham Boer War Diary 1: Report on RE work generally in South Africa 1 Dec 1900 to 1 Jan 1902, dated 13 May 1902.

[203] SANDF Archives, Photograph No. 771003366

[204] Brink, JN, *Recollections of a Boer Prisoner-of-war at Ceylon* (Jac Dusseau & Co, 1904). http://www.archive.org/stream/recollectionsab00bringoog/recollectionsab00bringoog_djvu.txt. Accessed 4 March 2014.

[205] Kearsey, *War Record of the York & Lancaster Regiment*, p182.

[206] Royal Engineers Chatham, Boer War Diary 1: Report on RE work generally in South Africa 1 Dec 1900 to 1 Jan 1902, dated 13 May 1902.

[207] Harris, JD, Major, 'Wire at War. Signal Communication in the South African War 1899-1902', *SAMHS Military History Journal*, Volume 1 (June 1998).

[208] *The Graphic* (August 10-15 1901).

[209] *Manual of Instruction in Army Signalling. Section III – Apparatus, And Method Of Using It* (1886).

[210] From a box of 89 stereoscopic cards entitled, 'South African War through the Stereoscope Pt 2 Volume 1' (Underwood & Underwood, 1900). Author's Collection.

[211] Harfield, A, *Early Signalling Equipment: The Heliograph: a Short History* (Blandford: The Royal Signals Museum, 1986).

[212] Royal Engineers Chatham, Boer War Diary 11, Army Telegraph Report 1900–1902.

[213] Austin, B, 'Wireless in the Boer War', *The Journal of the Royal Signals Institution*, Volume XXV No1 (Spring 2004).

[214] Baker, DC, 'Wireless Telegraphy during the Anglo-Boer War of 1899–1902', *The South African Military History Journal*, Volume 11, No.2 (December 1988).

[215] Ibid, p24.

[216] Memo No. 52, Adjutant-General, Maj-Gen WF Kelly, dated 3 December 1901.

[217] Moffett, EC, *Private, With the Eighth Division: A Souvenir Of The South African Campaign.* (Knapp, Drewett & Sons Ltd, 1903). http://www.archive.org/stream/witheighthdivis00moffgoog/witheighthdivis00moffgoog_djvu.txt. Accessed 4 March 2014.

[218] National Army Museum, London. Photograph NAM 1972-08-72-2-113 Negative No 134357.

[219] Royal Engineers Chatham, Boer War Diary 1: Report on RE work generally in South Africa 1 Dec 1900 to 1 Jan 1902, dated 13 May 1902.

[220] Honouring Black Dead in Boer War 26 September 2005, https://www.brandsouthafrica.com/south-africa-fast-facts/history-facts/war-graves-260905. Accessed 29 June 2020.

[221] Nkuna, N, 'Black involvement in the Anglo-Boer War, 1899–1902,' *Military History Journal*, Volume 11, No 3 (October 1999).

222 Mohlamme, JS, 'The Role of Black People in the Boer Republics during and in the Aftermath of the South African War', Doctoral thesis, University of Wisconsin-Madison, 1985, p99.
223 National Army Museum, London. Photograph NAM 1978-07-4-34, 'A contingent of native blockhouse guards', from an album of 44 photographs taken and collected by Colonel OM Harris DSO, Royal Artillery.
224 National Army Museum, London. Photograph NAM 1985-11-40-10 labelled Mounted 'Kaffir' Scouts 1900. One of 17 photographs by Lieutenant Charles Hare.
225 Van Heyningen, *Concentration Camps of the Anglo-Boer War*, p150.
226 http://www.sahistory.org.za/topic/black-concentration-camps-during-anglo-boer-war-2-1900-1902, 20 October 2014.
227 National Army Museum, London. Photograph NAM 1983-12-61-96.
228 The National Archives, London. PRO 30/57/22 Letter dated 9 Mar 1902 to Broderick. The Cape Boy Contingent was a body of coloured men who served with the British in the defence of Mafeking (*Dictionary of South African English on Historical Principles*), while 'Basters' refers to communities in the Northwest Cape who had traditionally done commando service to police the area (Giliomee and Mbenga, *New History of South Africa* (Cape Town: Tafelberg, 2007)).

CHAPTER 6: MOBILE BLOCKHOUSES

229 Steenkamp, W, *The Black Beret – Story of South Africa's Armour* (Draft Edition, 2001).
230 Schreier, KF Jr, 'Steam Tractors: The Power Behind First Motorized Armored Vehicles', *Farm Collector Magazine* (2001).
231 Cima, KH, *Reflections from the Bridge*. Baron Birch in conjunction with the Institute of Royal Engineers (1994), pp69-76.
232 Museum of English Rural Life, e-mail dated 18 February 2014. Engines delivered at different dates in 1900: 8894 – May 29th, 8895 – June 11th, 8896 – July 28th, 8897 – Aug 15th, 8898 & 8899 – Oct 19th, 8900 – Dec 31st.
233 https://tanks-encyclopedia.com/pre-ww1-uk-fowler-b5-armoured-road-train/. Accessed 30 June 2020.
234 The First Armored Motorized Military Vehicle, by engineered™, on May 11th, 2011. http://atomictoasters.com/2011/05/the-first-armored-motorized-military-vehicle/. Accessed 2 Feb 2014.
235 https://tanks-encyclopedia.com/pre-ww1-uk-fowler-b5-armoured-road-train/ 30 June 2020.
236 Royal Commission on the War in South Africa, Minutes of Evidence Taken Before The Royal Commission on the War in South Africa (London: HMSO Wyman And Sons, Limited, 1903).
237 *The Northern Echo*, 12 July 2010. http://www.thenorthernecho.co.uk/news/8266631.Engine_rolls_back_the_years/.
238 http://www.flickr.com/photos/preservedtransport/8127522061/in/photostream/.
239 Fowler Album TR FOW PH2/11 from the top plate, Nos B542, B540, B535 and B547.
240 Public Record Office (PRO Files) – The Kitchener Files. PRO 30/57/13. Letter S/19 Sir HS Rawlinson to Lord Kitchener suggesting blockhouse lines and mobile blockhouses on wagon dated 1901.
241 Durham University Library, Reference: SAD.628/119. Description: Steel forts on ox-drawn wagons, designed by E Wood to escort convoys.
242 Royal Engineers Chatham (Library and Archives), Boer War Diary 1: Report on RE work generally in South Africa 1 Dec 1900 to 1 Jan 1902, dated 13 May 1902.
243 Durham University Library and Collections SAD 628/129.

[244] Fuller, *The Last of the Gentleman's Wars*.
[245] Amery, *The Times History of The War in South Africa*, p145.
[246] Peters, *The Architecture of the Blockhouses in the Anglo-Boer War*, Part 2: 'Rice-Pattern', pp44-53.
[247] National Army Museum (NAM 1985-10-130-22). Photograph of elements of a prefabricated blockhouse on a flat bed wagon. From photograph album of 24 photographs taken by Lt Francis Hills Cobb, 1901. Associated with Royal Engineers, Boer War (1899–1902). Image 1018060.
[248] Letter from the RE Museum Chatham dated 11 April 1997. SA Military Museum file C.248.
[249] National Army Museum, London, Image No 1018059.
[250] Royal Engineers Chatham, Boer War Diary 5.
[251] Talana Museum reference: ABW 16201-1749, unknown source photocopy article titled 'Early Armour in the Union Defence Forces'.
[252] *The Graphic* (29 March 1902), p420.
[253] Bethell, 'The Blockhouse System in the Anglo-Boer War', Paper XII, 1904, Fig 2. Photograph courtesy of SANDF Photograph Archive No 771003048.
[254] www.kentishforum.co.uk. Accessed 20 April 2013.
[255] http://armoredcars-ww-one.blogspot.com/2012_10_01_archive.html - 30 October 2012.
[256] The tank's name was derived from that of its parent organisation, Allgemeines Kriegsdepartement, 7. Abteilung, Verkehrswesen. In German the tank was called Sturmpanzerwagen, (roughly 'assault armoured vehicle').
[257] By Hohum - Own work, CC BY 3.0, https://commons.wikimedia.org/w/index.php?curid=5013209.
[258] http://en.wikipedia.org/wiki/Bison_concrete_armoured_lorry. Accessed 5 May 2013.
[259] Original photograph, The War Office, Public domain out of Copyright https://en.wikipedia.org/wiki/Bison_concrete_armoured_lorry. Accessed 20 June 2020.

CHAPTER 7: LIFE IN THE BLOCKHOUSES
[260] 'British Regiments in 1881'. https://www.angloboerwar.com/other-information/16-other-information/294-british-regiments-in-1881. Accessed 20 May 2020.
[261] *Ballads of the Boer War: Selected from the Haversack of Sergeant J. Smith* by 'Coldstreamer' Grant Richards (London: 1902).
[262] In war games and exercises during the Cold War, friendly forces were termed the 'Blue Force' and the enemy the 'Red Force', hence the term for accidentally engaging your own force as 'Blue on Blue'.
[263] Fuller, *The Last of the Gentleman's Wars*, p114.
[264] 'Hawick Boer War Memorial'. http://warmemscot.s4.bizhat.com/warmemscot-ftopic101-15.html. Accessed 3 June 2007.
[265] 'From Beef and Chocolate to Daily Ration – British Rations in Transition 1870-1918'. http://17thdivision.tripod.com/rationsoftheageofempire/id5.html. Accessed 20 March 2015.
[266] Fuller, *The Last of the Gentleman's Wars*.
[267] Oatts, Lt.-Col. LB, DSO Late HLI, *Proud Heritage, The Story of the Highland Light Infantry, Volume Three, The Regular, Militia, Volunteer, T.A. and Service Battalions HLI, 1882–1918* (Glasgow: The University Press, 1961).
[268] Backhouse, JB, *With the Buffs in South Africa* (1903), Ray Westlake Military Books facsimile (1989), pp117-119.
[269] Singer, B, *Churchill Style: The Art of Being Winston Churchill* (Harry N Abrahams, 2012).
[270] King's College London Photograph Collection. Image: 4459/385. Stone built building with

corrugated roof, the floor has wooden boards there are 6 openings in the wall at roof level. A camp bed is set out along the wall on top of which is a pith helmet and other equipment.

[271] National Army Museum 1991-07-3. Printed orders for blockhouses, March 1902, issued to all ranks, the Queen's (Royal West Surrey Regiment).

[272] The War Museum Bloemfontein Reference 6648/10, copy of original orders.

[273] Colonel Godfrey commanded The 1st Battalion of the 25th of Foot (The Scottish Borderers) who were in theatre from January 1900 to June 1902. In September 1901 they were coming to the end of tour on the Mooi River line, when these orders were most likely written.

[274] 'The Orange River Colony', *The Times* (25 February 1902).

[275] Disease figures from *Blue Book* and *The Times History,* Volume VII.

[276] 'Second Boer War records database goes online'. https://www.bbc.com/news/10390469. Accessed 20 May 2020.

[277] A reference to Major-General Sir Leslie Rundle, British Divisional Commander.

[278] A reference to Joseph Chamberlain, Colonial Secretary under Lord Salisbury.

[279] Welch, JC. 'The Blockhouse Line', *Journal of the Society for Army Historical Research*, Volume 83, No 334 (2005), pp93-109. JSTOR, www.jstor.org/stable/44231167. Accessed 12 May 2020.

[280] *The Graphic* (22 March, 1902), p396.

[281] Buttery, David, 'A Toff's Life in the Blockhouse', *Soldiers of the Queen*, Issue 103 (December 2000).

CHAPTER 8: EFFECTIVENESS OF THE BLOCKHOUSE SYSTEM

[282] De Wet, *Three Years' War*, Chapter 31.

[283] Zurnamer, Major BA, 'The State of the Railways in South Africa during the Anglo-Boer War 1899–1902', *South African Journal of Military Studies,* Volume 16, No 4 (1986).

[284] Ibid, p28.

[285] Royal Engineers Chatham, Boer War Diary 9, 23rd (Field) Company RE Notes 1899-1902, dated 31 May 1902.

[286] Ibid, p14.

[287] The National Archives of the UK (TNA), WO 78/5450, 'E' Line of Communications: plan of railway showing positions of blockhouses etc. Dated 1902

[288] Frederick, Major-General Sir Maurice, K.C.B. with a staff of Officers, *History of the War in South Africa 1899–1902,* Volumes 1-4 (London: Hurst and Blackett Ltd, 1906).

[289] 'The British Command Papers, commonly known as 'Blue Books' from their covers, are collections of official papers presented to Parliament. They contain much invaluable information and are standard reading on the camps, but readers should never forget that they are documents presented by the party in power and are intended to reflect well on their actions. A careful comparison of the camp Blue Books with the original documents demonstrates, however, that they were accurate and full, with remarkably little censorship. As far as they go, they can be used confidently.' University of Cape Town, British Concentration Camps of the South African War archive note.

[290] McLeod, 'The Psychological Impact of Guerrilla Warfare on the Boer Forces During the Anglo-Boer War', University of Pretoria Thesis (2004), p89.

[291] Davies, K, 'Armoured Trains', *War Monthly*, No18 (September 1975), pp43-46, and also Royal Engineers Institute, *Detailed History of the Railways of the South African War 1899–1902* pp248-50.

[292] Blue Book ,Volume 1, Appendix IV, p459.

²⁹³ Ibid, p459.
²⁹⁴ http://www.angloboerwar.com/other-information/16-other-information/1845-cost-of-the-war 5. Accessed November 2014.
²⁹⁵ Bethell, 'The Blockhouse System in The South African War' (1904), p289.
²⁹⁶ PRO 30/57/22 The Kitchener Papers, 13 Dec 1901 Letter from Lord Kitchener to St John Brodrick, Secretary of War.
²⁹⁷ De Wet, *Three Years' War*, Chapter 31.
²⁹⁸ Ibid, Chapter 31.
²⁹⁹ De Wet, *Three Years' War*, p268.
³⁰⁰ Atwood, *Roberts & Kitchener*, p248.
³⁰¹ Bryant, *26.2 – The Incredible True Story of the Three Men Who Shaped the London Marathon*, Chapter 4.
³⁰² Doyle, AC, *The Great Boer War*, Project Gutenberg, www.gutenberg.org (2009).
³⁰³ King's College London, The Hamilton Diaries, 29 Nov 1901 Letter to Lord Roberts.
³⁰⁴ 'De Wet and the Blockhouse Lines', *The Times*, 8th December 1902.
³⁰⁵ Royal Engineers Chatham, Boer War Diary 8.
³⁰⁶ Ibid.
³⁰⁷ Royal Commission on the War in South Africa, Minutes of Evidence Taken Before the Royal Commission on the War in South Africa (London: HMSO Wyman And Sons, Limited, 1903).
³⁰⁸ Reitz, *On Commando*.

CHAPTER 9: THE ARCHITECTS OF THE BLOCKHOUSE SYSTEM
³⁰⁹ Lord Roberts, Courtesy National Portrait Gallery (P1700(31c)).
³¹⁰ Bethell, *The Blockhouse System*, p277.
³¹¹ Wessels, André, 'Frederick Roberts', in Corvi and Beckett, *Victoria's Generals,* Chapter 7, p167.
³¹² Farwell, *The Great Boer War*, supra note 31, p352.
³¹³ Corvi & Beckett, *Victoria's Generals*, p196.
³¹⁴ Lord Kitchener by Alexander Bassano. Courtesy National Portrait Gallery (x96309).
³¹⁵ Corvi & Beckett, *Victoria's Generals*, p196.
³¹⁶ Samson, A, 'Kitchener of Khartoum', SAMHS Evening Lecture, December 2012.
³¹⁷ Sudan Railway Corporation, Facts & Figures Yearbook 2007, downloaded 10 August 2020.
³¹⁸ Steevens, *With Kitchener to Khartum*, p22.
³¹⁹ Corvi & Beckett, *Victoria's Generals*, p209
³²⁰ The Van Plettenburg Historical Society, *The Battle Of Omdurman,* 11 May 2014, https://www.pletthistory.org/talks/the-battle-of-omdurman/ Accessed 10 August 2020
³²¹ Atwood, *Roberts and Kitchener in South Africa 1900-1902* (2011), p57.
³²² Laws, *Who Killed Kitchener?* p67.
³²³ Corvi & Beckett, *Victoria's Generals*, p199.
³²⁴ Atwood, *Roberts and Kitchener in South Africa 1900-1902* (2011), p206.
³²⁵ Valiunas, *Churchill's Military Histories: A Rhetorical Study*, p39.
³²⁶ Wilson, *A Life of Sir Henry Campbell-Bannerman*, p349.
³²⁷ Atwood, *The Life of Marshall Lord Roberts*, p222.
³²⁸ Farwell, *The Great Boer War*, p358.
³²⁹ The National Archives, Kitchener Papers PRO30/57/21 dated 5 Nov 1901, telegram K to Lord Roberts.
³³⁰ The National Archives Kew, PRO 30/57/22 13 Dec 1901, Letter from Kitchener to Brodrick.

331 King's College London Website, https://kingscollections.org/catalogues/lhcma/collection/h/ha30-001/
332 Liddell Hart Centre for Military Archives (King's College London Archives), The Hamilton Diaries, Hamilton 2/3/1 Letter dated 13 November 1901.
333 Atwood, *Roberts & Kitchener*, p244.
334 Ibid, p248.
335 Macdonald, Wade *et al, HMS HAMPSHIRE 100 Survey 2016, Survey Report*, p32.
336 Bethell, *The Blockhouse System*, p278.
337 Lt Col Baden Powell, by FH Hart. Courtesy National Portrait Gallery (82301).
338 'Who Invented the Block-House System?' *Pretoriana, Magazine of the Old Pretoria Society*, No 73 (1977), pp15-16.
339 Ibid, p16.
340 Ibid, p17.
341 Ibid, p21.
342 Ibid, p22.
343 http://www.wishful-thinking.org.uk/genuki/GLS/Newent/MIs.html. Accessed 25 July 2009.
344 *British Medical Journal* (25 August 1898), p578.
345 Obituary Major-General Sir Elliott Wood, *The Times,* 7 September 1906.
346 Wood, *Life and Adventure in Peace and War*.
347 Ibid.
348 Bethell, 'The Blockhouse System in The South African War', Plate XIV.
349 The Medjidieh was an Ottoman order and decoration instituted in 1852 by Sultan Abdul Medjid as both a civil and a military award. There were five classes in the order, and the decoration consisted of a silver sun with seven threefold rays, which alternate with the crescent and the star.
350 The Khedive of Egypt presented a bronze star to all the officers who served to engage in the suppression of the rebellion in Egypt in 1882, which is always presented in conjunction with the British Egypt Star.
351 http://www.englandspastforeveryone.org.uk/webdav/pdfconverter?asset=1678. Accessed 25th July 2009.
352 Wood, *Life and Adventure in Peace and War*, p268.
353 'The West Redoubt, National Army Museum' Postcard by Charles Clarke (Haywards Heath) Ltd; the P'tit Blockhouse (Wikipedia, https://commons.wikimedia.org/wiki/Category:Fortin_du_P%27tit_Sault#/media/File:Fortin_du_P'tit_Sault_2014.JPG).'
354 Wood, *Life and Adventure in Peace and War*, p156.
355 Ibid.
356 'Pontoon', http://www.1911encyclopedia.org/Pontoon. Accessed 18 July 2009.
357 Gooch, J, *The Boer War: Direction, Experience and Image* (Cass Series – Military History & Politics) (2000), p52.
358 'The Boer Story of the War', *The Times* (1 December 1902).
359 *London Gazette*, Issue 27282 (8 February 1901), p845
360 *London Gazette*, Issue 27305 (16 April 1901), p2601.
361 Wood, *Life and Adventure in Peace and War,* p269.
362 Ibid, p269.
363 Royal Engineers Chatham Library and Archives Boer War Diary 1. Summaries of Engineer Work sent to the War Office, 4 June 1901, CRE Army Elliot Wood.

[364] Wood, *Life and Adventure in Peace and War*, p269.
[365] *London Gazette*, Issue 27306 (10 April 1901), p2695.
[366] *London Gazette*, Issue 27459 (29 July 1902), p4836.
[367] *London Gazette*, Issue 27490 (31 October 1902), p6897.
[368] Published by E Arnold & Co (London, 1924).
[369] Obituary, *The Times* (13 August 1929).
[370] *Thompson's Irish Who's Who.*
[371] *London Gazette*, Issue 24581 (14 May 1878), p3045
[372] *London Gazette*, Issue 25815 (11 May 1888), p2698
[373] *London Gazette*, Issue 25955 (19 July 1889), p3896.
[374] *London Gazette*, Issue 26740 (19 May 1896), p2989.
[375] General SR Rice, National Portrait Gallery (x80750).
[376] *Cape Times* Weekly Edition (5 July 1899).
[377] Rickard, J (5 February 2007), 'Battle of Lombard's Kop, 30 October 1899'. http://www.historyofwar.org/articles/battles_lombard_kop.html. Accessed 28 August 2009.
[378] *London Gazette* (8 February 1901), p917.
[379] http://www.angloboerwar.com/Other/army_mentions.htm#LADYSMITH
[380] Wilson, HW, *After Pretoria: The Guerrilla War*, p891.
[381] *London Gazette* (26 June 1902), p4194.
[382] Free-BMD Website (Q3 1a p288) in Kensington, London.
[383] *London Gazette*, Issue 28182 (2 October 1908), p7114.
[384] *London Gazette*, Issue 28879 (25 August 1914), p668.
[385] Supplement to the *London Gazette*, Issue 28724 (3 June 1913), p3904.
[386] *London Gazette*, Issue 29177 (1 June 1915), p5214.
[387] https://en.wikipedia.org/wiki/Spring_R._Rice. Accessed 10 November 2014.
[388] Supplement to the *London Gazette*, Issue 30111 (4 June 1917), p5458. The Most Distinguished Order of Saint Michael and Saint George is a British order of chivalry founded on 28 April 1818 by George, Prince Regent (later George IV) whilst he was acting as Prince Regent for his father, George III. It is named in honour of two military saints, St Michael and St George. Awarded as the Knight Commander and may use the prefix 'Sir'.
[389] Supplement to the *London Gazette*, Issue 30202 (26 July 1917), p7590.
[390] Supplement to the *London Gazette*, Issue 30568 (11 March 1918), p3097.

CHAPTER 10: THE TRANSITION TO PEACE
[391] Balfour, R, 'Draft Lecture', 9, *Balfour Papers* (1944), the story of which is now immortalised in the Hollywood film, 'The Monuments Men'.
[392] Gillings, K, 'The Aftermath of the Anglo-Boer War', SAMHS presentation, 26 August 2008.
[393] Ibid.
[394] *The Standard*, Krugersdorp (9 May 1903).
[395] Maturin, Mrs Fred, *Petticoat Pilgrims on Trek*, Eveleigh Nash (London, 1909).
[396] Gillings, K, 'The Aftermath of the Anglo-Boer War', SAMHS presentation, 26 August 2008.
[397] Tomlinson, *Britain's Last Castles*.

INDEX

1st Corps, Royal Engineers, 250
1st King's Own Scottish Borderers (KOSB), 194-195
1st Sudanese Brigade, 235
2nd East Kent Regiment (The Buffs), 199
3rd Leicestershire Regiment, 209
4th (Royal Irish) Dragoon Guards, 14, 192
4th Battalion The Rifle Brigade, 209
5th Company Bombay Sappers
6th Infantry Regiments (United States), 33
7th (Field) Company RE, 133
13th Infantry Regiments (United States), 33
15th (The King's) Hussars, 14, 192
17th (Field) Company RE, 29, 133, 191, 246
20th (Fortress) Company RE, 133
23rd (Field) Company RE, 100, 118, 120, 131-132, 191, 249, 251-253
24th Infantry Regiment (United States)
26th (Field) Company RE, 191
38th (Field) Company RE, 180, 191
45th (Fortress) Company RE, 176
47th (Field) Company RE, 133, 185, 191

A

Abraham's Kraal, 59, 265-266
Abu Hamed, 81
Abyssinia, 229
Afghanistan, 27-28, 88-89, 98, 230, 233
African auxiliaries, 16
African drivers, 135
African guard, 196, 200, 202
African Ring, 152
African people, 16, 69-70, 149, 169, 171-173, 194, 196, 221, 226, 257
Afrikaners, 34, 39, 49, 237, 281
Agterryers, 171, 276
Alarm devices, 161
Alarm guns, 160, 162
Aldershot, 248, 254-255
Alicedale Junction, 265
Aliwal North, 155, 258, 266
Aliwal North, blockhouse, 103, 110-111
Allison's Kop, 269

Ambush, 31, 46, 89
Amelia, 136
America, South African town, 83
American Civil War, 22, 64, 176
American Revolutionary War, 20
Amsterdam, South African town, 269
Anglo-Egyptian War, 28
Anglo-Zulu War, 13, 36, 95-96, 135
Anti-mobility ditch, 159
Armour plate, 30, 178
Armoured train, 80, 83-86, 144, 161, 168, 183-184, 219
Armoured train telegraph, 168
Armoured wagon, 175, 182
Armstrong 9-pounder, 41
Army Signalling Manual, 163
Aroostook Wars, 247
Artillery, 18-20, 24, 27, 41, 44, 46, 59, 78, 80, 83, 93, 97-98, 121, 128, 151, 189, 213, 216, 229, 232, 246, 277, 283-285
Arundel, 42
Ashanti Ring, 229
Aties Fort, 106
Attacks on blockhouse, 225

B

Baden-Powell, Colonel Robert 46, 57, 140, 241
Bag or trap, 75, 246, 276
Bakleikraal, fort, 106
Balloon, 167
Barbed wire, 30-32, 52, 100, 119, 126, 135, 154, 157-158, 161, 203, 205, 209, 225, 271-272
Barberton, 100, 212, 265
Barberton, blockhouse, 130, 144
Barkley East, 42
Barracks, 22, 94, 98, 133, 188, 245, 254, 259, 278
Barwell, MS, Miss, 173
Bastards, Cape, 174
Battle of Bergendal, 65
Battle of Cambrai, 189

Battle of Colenso, 43
Battle of Elandsfontein, 96
Battle of Groenkop, 73, 134, 239
Battle of Lombard's Kop, 251
Battle of Magersfontein, 43-44
Battle of Nicholson's Nek, 251
Battle of O'OKiep, 145
Battle of Omdurman, 235
Battle of Paardeberg, 44-45, 236
Battle of Poplar Grove, 45
Battle of Sanna's Post, 46
Battle of Spion Kop, 55
Battle of Stormberg, 43
Battle of Tweebosch, 239
Beaufort West, 160, 265
Beaufort West, blockhouse, 62
Beer, 200, 209
Belfast, 253
Bellville Junction, 265
Benneyworth, Garth Dr, 173
Benson's Column, 238
Benson, Lieutenant-Colonel GE, 78
Berber, 29, 246
Berlin, 188
Bethell, Colonel EH, 28, 105, 127, 135, 137, 232, 241, 246
Bethlehem, 267-269
Bethulie, 212, 265-266
Bezuidenhout, Piet, 36
Bimaru Heights, 27
Bison armoured vehicle, 189-190
Black
    Concentration camps, 50, 65, 67-69, 173
    Families, 50, 256
    Native workers, 13-14, 35, 50, 65, 69, 71-73, 78, 88, 103, 127, 134-136, 141, 150-151, 154, 171, 174, 194, 220, 222, 256, 259, 261
Black Week, 43, 53, 143, 230
Blast wall, 122, 124
Blockhead Policy, 224, 248
Blockhouse
    Advantages of different designs, 13, 49, 77, 129-130
    Cost effectiveness, 62, 102, 125, 128, 220-222, 249, 257
    Cross-country lines, 58, 72, 74, 131, 134-136, 139-140, 165-166, 217, 244, 266-269
    Crossing the line, 63, 86, 161, 217
    Disadvantages of different designs, 129-130, 187
    Factory locations, 132-133
    Kits, 71, 128, 131-135, 182-183, 249
    Line distances, 139
    Mass-production of, 63, 120, 123-124, 126, 128-129, 131, 185, 214, 252
    Materials & Labour used to build, 271
    Mobile blockhouses, 175, 183, 186-189, 191
    Moveable line of blockhouses, 182
    Orders for occupation, 201
    Prefabrication, 17, 97, 100, 122, 125, 128, 130, 133, 135, 189
Blockhouse patterns
    Brindisi pattern, 112
    Burnaby pattern, 125, 128, 259-260
    Elliot Wood pattern, 60, 62-63, 102, 128-129, 187, 213, 216, 245-246
    Hexagonal, 110, 115-116, 142, 156, 284
    Ladysmith pattern, 125-126
    Octagonal, 61, 107-109, 122-125, 127, 130, 132, 201, 214, 252, 260, 283
    Pepper-pot, 128, 136
    Rectangular, 111
    Round, 106-107, 124-125, 127, 130
    Square, 107-108, 112, 114-115, 206, 235
    Umbrella roof pattern, 110, 124
    Vereeniging pattern, 112
Bloemfontein
    Conference 1899, 41
    Defence of, 58-59, 133
    Strategic objective, 42, 45, 53, 140, 215, 231-232
    Town defences, 133, 266
Bloemhof, 269
Boer
    Collaborators, 78
    Homesteads, 256
    Militarisation, 41, 167
    Modernisation, 41
    Posturing, 42
    Republics, 12, 35, 39-40, 52-53, 57, 68, 88, 145, 151, 256

Boetzelaer Battery, 19
Bois Blanc Boblo Island, 23
Bombay Sappers, 28
Boshof, 267, 269
Botha's Pass, 137, 268
Botha, General Louis, 46, 48, 70, 222, 224
Bothaville, 267, 269
Bovington Tank Museum, 190
Bowman Radio System, 169
Brabant's Horse, 231
Brandt, Johanna, 215
Brandwater Basin, 47, 112, 114
Breastwork, 278
Bridges, 82, 103-104, 107, 112, 123, 192, 207, 211-213, 216, 248, 264, 267-268
British
    Army, 13-14, 16, 35-36, 38-39, 41-42, 46, 52, 55, 68, 79, 98, 140, 151, 166, 168-169, 171-172, 192, 200, 232-233, 240, 251, 259
    Colonial, 18, 49, 141, 194
    Crown, 25, 88
    Empire, 12-13, 19, 34-35, 43, 81, 87-88, 151, 176, 229, 232, 283
    Expeditionary Force, 254-255
    Legion, 250
Brodrick, William St John Freemantle, 237-238
Brother Boer, 193
Brugspruit, 85, 267
Buffelsvlei, 268
Buller, General Sir Redvers, 42-43, 45, 48, 58, 230, 232, 248, 252
Bully beef, 161, 197, 209
Bunker, 153, 278
Burgersdorp, blockhouse, 103-104, 248
Burghers, 37, 46, 69, 85-86, 222, 244
Burney's Column, Colonel, 162
Burning farms, 237
Burrell, Charles & Sons, 176
Bushmen Relic Protection Act, 261
Bushveld Carabineers, 98
Buston, Lieutenant-Colonel PT, 133
Buttress, 278, 285

C
Cable, 92, 114, 166, 277
Caisson, 127, 273, 278
Calvinia, 71, 269
Camberley, 152
Cammell, Charles & Co Ltd, 176, 179
Camouflage, 43, 130, 151, 183, 188-189
Campos, General AM, 29
Canada, 20, 23, 81, 243, 247
Canadian Mounted Rifles, 150
Cannon, 19, 21-22, 276-277
Cape Afrikaners, 49
Cape Boys, 174
Cape
    Colony, 34-35, 40, 42, 49, 52-53, 57, 61, 71, 95, 103, 106, 116, 125, 139, 143-144, 150, 174, 205, 211-213, 225, 232
    Dutch, 61, 143
    Frontier Wars, 13, 93
Cape Town, 16, 19, 42, 52-53, 62, 71, 92-93, 103, 116-117, 131, 133, 140, 143, 166-167, 206, 243, 245, 250, 252, 265
Cardwell Reforms, 151
Carnarvon, 71, 240, 268-269
Carnarvon, blockhouse, 63, 71
Carolina, 268
Castle of Good Hope, 16, 92-93, 167, 262
Casualty, 218, 226
Cattle, 35, 64, 74-75, 78, 80, 86, 157, 170, 181, 205, 219, 242, 258, 268
Cattle Pass, 268
Causes of death, 204
Cederberg Range, 140
Cement, 114
Centenary, 17, 88, 262-263
Central Park, 21
Cetshwayo, King, 36
Ceylon, 161
Chain-of-command, 50, 75, 224, 236
Chamberlain, Joseph, 206, 244
Chappe, Claude, 94
Chatham, Kent, 90, 100, 188, 234, 245, 254-255, 270
Chatham House, 90
Chatham Memorial Arch, 188, 254-255
Chelmsford, General Lord Thesiger, 68, 95-96
Chocolate, 197-198, 207
Chunespoort, 267

Citadel, 279
Civilian administration,, 68, 70
Clanwilliam, 68-269
Clanwilliam, blockhouse, 71
Clegg, Right Honourable Nick, 90
Colenso, 42-43, 212, 230
Colesberg, 41-42
Colleton, Colonel Sir Robert, 180, 191
Colley, Major-General Sir George Pomeroy, 37-38
Colonial Defence Force, 143
Columns, 49, 51, 57, 66, 70-73, 76-78, 80, 86, 90, 95, 134, 149-150, 166, 172, 175, 177, 180, 192, 217, 220, 232, 238-240
Commando, 13, 34, 43-44, 50, 54-55, 64-66, 70, 72-78, 106, 143-144, 162, 171, 173, 194, 215, 217, 222-223, 238-239, 241
Commando Drift, 269
Commando Drift (Valsch River) Extension, 269
Communication, 15, 49, 52-53, 56, 61, 63, 84, 94, 99, 123, 140, 143, 148, 162-166, 168, 190, 209, 214, 216, 235, 238, 253
Concentration camp, 14, 68-69, 221, 225
Concrete, 102-103, 114, 189-190, 275, 284
Conventional conflict, 42, 44, 46-47, 54, 232, 237
Conversions of existing structures, 117
Corrugated iron, 17, 63, 89, 99, 105, 114, 116, 119-120, 122-124, 128-129, 132, 137, 149, 162, 184-185, 189, 199, 204, 207-208, 257-258, 262
Corsica, Spain, 20
Cost effectiveness, 128, 220
Council of war, 45, 282
Counter-mobility, 76-77, 217, 231, 233, 238, 240, 274
Counterinsurgency (COIN), 13, 87
Cow Tower, England, 18
Cradock, 265
Craig, Major-General Sir James, 92-93
Creusot, 41
Crimean War, 35, 151
Crockett Blockhouse, America, 22
Crockett, Colonel Walter, 22
Cronje, General Piet, 44
Cuba

War of Independence, 12, 29-33, 65, 67, 98, 284
Jungle, 30-31
Cyclist Company, 144

D
Danaher, Trooper John VC, 96
Danielskuil, fort, 106, 116
Darley, Sir Frederick, 41
Davis, Richard Harding, 31
Dayton Mobile Blockhouse, 189
De Aar, 53, 160, 165, 167, 169, 265
De Beer's Pass, 268
De La Rey, General Koos, 46, 184
De Lange Drift, 266
De Wet, General Christiaan, 46-47, 49, 65, 72-77, 79, 84, 103, 124, 134, 161, 185, 209, 211, 217, 222-224, 240, 248, 256
Delagoa Bay, 53, 55, 123, 167-168, 214, 225
Delhi, 27
Denial of movement, 71, 216
Department of Native Refugees, 69
Dervishes, 235
Detroit River, 23
Devil's Peak, 92-93
Devonshire Regiment, 142, 158
Diamonds, 36, 39
Digby-Jones, Lieutenant RJT VC, 251
Dispatches, 165, 229, 237-238, 246, 248, 252-253, 255
Doctrine, 14, 77, 79, 89-90, 280
Doyle, Sir Arthur Conan, 224
Drakensberg, 37
Drury, Curate William, 206
Dummy, (Sentry), 193
Dundee, 138, 267, 269
Dunkirk, 189
Durban, 42, 52-53, 71, 131, 158, 186
Dutch, 12, 34-35, 61, 88, 93, 143
Dutch East India Company, 34, 93
Dwyka River, 62
Dynamite, 55, 85
Dysentery, 173, 204-205

E
Earthwork, 26, 142, 278, 280, 285
East Fort, Pretoria, 60, 123

Echelons, 76, 169, 172
Eerste River, 265
Egypt, 28-29, 81, 100, 118, 123, 212, 234, 246, 250
Egyptian Army, 81, 234-235
El Caney, Cuba, 33
Elands River, 267
Elandsfontein, 71, 133
Elphinstone, Rear-Admiral George, 20
Enclaves, 32, 71
Encrypted messages, 166
Enfilade fire, 22, 28, 107, 130, 279-280
English Heritage, 18
Enslin, 167
Enteric, 45, 204, 226
Ermelo, 222, 258, 268-269
Erste Fabriken, 266
Eshowe, 96

F
Fairfield Blockhouse, America, 23
Fashoda, 235
Fawcett, Millicent Garrett, 119
Fences, wire, 126, 153-154, 157-162, 209-210, 271, 274
Ficksburg (Brindisi), 268
Fieldcraft, 13, 43, 231
First World War, 91, 152, 168, 178, 186, 189-190, 233, 240, 254, 263
Flanking fire, 104, 108, 110, 276
Fort, 16, 19, 21, 26, 106, 116, 142-145, 242-243, 280
    Aties, 106
    Campbell, 113-114
    Commeline, 96
    Daspoortrand, 97
    Edward, 98
    Handub, Egypt, 29
    Hendrina, 97-98
    Jackson, 142
    Klapperkop, 97
    Knysna, 145
    Pearson, 96
    Peddie, 94
    Pitt, America, 20
    Roberts, 27
    Royal, 96
    Schanskop, 97
    Selwyn, 94
    Tenedos, 96
    Tullichewan, 96
    Wonderboompoort, 97
Fortified house, 56, 129, 281
Fortress, 16, 33, 279, 281
Fortress Artillery Corps, 41
Fourteen Streams, 212
Fowler B5 traction engine, 176-178
Fowler, John & Co, 176, 178-179
Franco-Prussian War, 64
Frankfort, 266, 268
French Revolution, 94
French Revolutionary War, 19
French, General Sir John, 44, 173-174
Frontier forts, 16, 118
Frontier War, 94-95
Fuller, Major-General JFC, 149, 196, 199

G
Gabion, 99-100, 281
Gates, John Warne, 157
Gatling gun, 33
Germany, 36, 97
Germiston, 133
Girouard, Major Edward Percy Cranwill, 79, 81-82, 212
Gladstone, Prime Minister William, 38
Glen Siding, 265
Godfrey, Colonel JW DSO, 204
Gold Coast, 78
Gold mines, 60, 69
Goldfields, 46, 127
Gone native, 172
Gordon, Major-General Charles George CB, 234-235
Graafwater Farm, blockhouse, 71
Great Trek, 35-36, 93
Greylingstad, 266-267
Groenewald, Commandant, 85
Groot Nyl Bridge, 207
Groot Olifants River, 124
Groot Oliphants, 267
Guerrilla, 12, 30, 46, 52, 87, 89
Guerrilla war, 48-49, 61, 65, 90, 143
Gun Hill, 124, 210

H

Haggitt, Major ED, 58
Hamilton, Colonel Ian, 57, 65, 75-76, 139, 149, 181, 224, 238
Hamilton, General Bruce, 57
Hanover Road, 265
Harrismith, 103, 145, 248, 260, 267, 269
Harrismith, blockhouse, 62
Hartbeespoort, 114
Headquarters, 26, 96, 108, 118, 133, 149, 165, 167, 169, 230, 248-249, 254-255
Heidelberg, 37, 266
Heilbron, 83, 266
Hekpoort, blockhouse, 114, 125
Heliograph, 162-164, 166
Hensoppers, 73, 281
Hereford Voluntary Motor Corps, 250
Hickie, Colonel James Francis, 184
Highland Brigade, 181
Highveld, 17, 172, 204, 240, 259
Hobhouse, Emily, 70, 119
Home Guard, 143, 189
Hoopstad, 268-269
Hopetown, 36
Hostages, 64
Hout Nek, 137
Howitzer, 22, 60, 123, 179
Howitzer Redoubt, 60, 123
Hugel, Captain von, 136
Huggan, Private John, 194-195
Human component, 148-149
Human shields, 64
Hybrid, blockhouses, 15, 101, 115

I

Imperial Light Horse, 78
Imperial Military Railway, 150, 178
Imperial Yeomanry, pay, 221
Indian Mutiny, 25, 229
Indian Ring, 229
Indwe Extension, 266
Inspector General of Fortifications, 245
Intelligence, 31, 49, 72-73, 76-78, 80, 90, 137, 141, 151, 210, 222, 224, 235, 240
Interest-free loans, 256
Interned, 50, 68, 150, 173
Isandlwana, 36

J

Jack Hindon, 55, 213
Jacobsdal, 58-59, 106, 265
Jacobsdal, blockhouse, 111
Jameson Raid, 40, 55, 96
Johannesburg, 27, 40, 46, 53, 58, 60, 71, 97, 107, 112-113, 123, 125, 133, 140, 180, 188, 191, 215, 248, 252, 260
Johannesburg Fort, 41
Johannesburg, town defences, 266
Johnston's Redoubt, blockhouse, 60, 100, 115, 123
Johnston, Andrew, 115
Jones, Sergeant-Major J, 99-100
Jordan Siding, 196
Joubert, General Piet, 37, 97, 252

K

Kaalfontein, 268
Kaalfontein, blockhouse, 107
Kaap de Goede Hoop, 34
Kaapmuiden, 118
Kabul, 27, 229
Kandahar, 65, 229, 233
Kaapmuiden Extension, 265
Karee Station, 58
Keep, 75, 82, 96, 106, 126, 150, 153, 170, 197, 199, 212-213, 236, 258, 282-283
Kekewich, Colonel Robert, 184, 189
Kelly, Major-General WF, 169
Kempton Park, 107, 260
Kennedy, Captain JNC, 166-168
Key point, 57, 103, 200
Key towns, 48, 54, 58, 60, 140, 215
Khartoum, 65, 82, 234-235
Khazi, (slang toilet), 205, 282
Khedive's Star Medal, 246
Kimberley, 39, 41-42, 44, 53, 133, 177, 191, 212, 252, 269
King Edward VII, 254
King's Blockhouse, 92-93
King's College London, 244
King's Royal Rifle Corps, 152
Kipling, Rudyard, 163, 192
Kitchener Ring, 180
Kitchener's Horse, 231
Kitchener, Lord Horatio,

Armoured trains, 79
Blockhouse strategy and defences, 32, 56, 58, 61, 73, 77, 99, 123-124, 128, 216, 244, 248-249
Chasing De Wet 47
Chief-of-Staff, 43
Concentration camps, 66-70, 232
Drives, New Model, 75-76, 240
Egypt, 28, 81, 231, 234-235
Force Levels, 49
Girouard, Percy Cranwill, 81-82
Gone native, 172
Guerrilla war, 6, 228, 237
Jack Hindon remarks, 55
Moveable blockhouse line, 181-183
Paardeberg, Battle of, 236
Reluctance in command, 238-240
Reports to Parliament, 210
Scorched earth policy, farm burning, 50-51, 65, 66, 70, 223, 237
Strategy, 'on-the-fly', 91
Revenge for Gordon, 236
Use of blacks, 150-151, 172
Klapmuts Station, 265
Klein Karoo, 140
Klein Olifants River, 61, 123
Klerksdorp, 60, 184, 191, 266-268
Klip River, 111-112
Knapdaar, 209
Komatipoort, 48, 54, 56, 60, 99-100, 117-120, 123-124, 199, 212-214, 232, 246, 266
Kommando, 13, 103, 114, 140
Kommando Nek, blockhouse, 103, 114, 140
Koopmans River, 266
Kraaifontein Junction, 265
Krakedouw, fort, 106
Krijgsraad, 45
Kritzinger, General Pieter H, 143-144, 150
Krom River, blockhouse, 102
Kroonstad, 45, 74-75, 82-83, 210, 266, 268
Kruger, President Paul, 40-41, 48-49, 167
Krugersdorp, 56, 113, 258
Krupp, artillery pieces, 41

L
Laager, 73, 79, 222, 248, 282
Lace Diamond Mine, 266
Ladybrand, 58, 222, 267
Ladysmith, 42-43, 45, 53, 120, 125-126, 230, 251-253, 260, 268
Ladysmith, town defences, 269
Laing's Nek, 37-38
Laingsburg, blockhouse, 62, 102
Lambert's Bay, 71, 125, 240, 268
Landed gentry, 152
Landsdowne, Lord Henry, 230
Langeburg Range, 140
Langkrans Nek (Botha's Pass), 268
Langton, Hugh Hornby, 23
Learoyd, Lieutenant CD, 133
Lee-Metford, 85, 179
Leeds Militia, 24
Leeu Gamka, blockhouse, 62
Leeuw Spruit, 267
Leeuwspoort, blockhouse, 156
Leslie, Lieutenant-Colonel, 210
Levelling gauge, 127
Liberals, 49, 70, 151
Lichtenburg, 133, 269
Liddell Hart Centre, 239
Lightning strikes, 204
Lindley, 74-75, 125, 268-269
Line of Communication, 63, 214
Lockmaster's Blockhouse, Canada, 24
London Convention 1884, 40
Long Tom, artillery piece, 41
Loophole, 119, 122, 137, 202, 258, 276, 282
Looted, 55, 64
Loreto Convent, Pretoria, 96
Lourenço Marques, 48, 100, 118, 123-124, 168, 214
Lushai Campaign, 229
Lydenburg Extension, 268
Lyndenburg, 253

M
Machadodorp, 48, 268
Machavie, 266
Machine gun, 104, 152, 156, 284
Madras Army, 232
Mafeking, 41-42, 46, 53, 133, 191, 252, 269
Magaliesberg, 60, 113-114, 140
Mahdism, 82, 235
Majuba Hill, 38

Makhado (Louis Trichardt), 97
Malan, 49
Malmsbury Extension, 265
Malnutrition, 67
Malta, 245, 248
Manhattan, 21
Maori Wars, 25-26
Marconi, wireless radio, 12, 166-169
Maretzani, 269
Maritz, Manie, 49
Marksmanship, 39, 90
Martello Tower, 16, 19-20, 93, 283
Martini-Henry, 41, 55
Mary Dalrymple Scannell, 254
Mathews, Charles Bernard, 190
Mauser, 41, 179
Maxim, 41, 83, 152, 156
Maxwell, Lieutenant-Colonel RC, 103, 133
McLaren, J&H & Co, 176
Medieval, 28-29, 113, 126, 149, 199
Mellon, Lieutenant C, 135
Merriman, blockhouse, 62, 103
Methods of barbarism, 237
Methuen, General Lord Paul, 167, 239
Middelberg, 61, 71, 191
Militia, 24, 26, 221
Milner, Lord Alfred, 41, 57, 65, 140, 143, 231, 236-238, 256
Milner's Kindergarten, 256
Mine Resistant Ambush Protected (MRAP), 89
Mining, 40, 60, 116, 251
Mobility, 46, 49, 76-77, 149, 158, 175, 187, 211, 217, 231, 233, 238, 240, 274
Modder River, 59, 62, 103, 167, 212, 250, 266
Modderfontein, 266
Montagu, 144-145
Montagu District Mounted Troop, 145
Montagu Town Guard, 145
Mooi River, 268
Moony, Major BE, 48, 90
Morale, 37, 45, 66, 70, 76, 90, 144, 150, 196, 199
Morant, Lieutenant 'Breaker', 98
Morris, Colonel (CRE) Cape Colony, 106
Morse Code, 162, 164, 284

Motte & Bailey, 126, 282-283
Mournful Monday, 252
Mr Fry, 241, 243-244
Muizenberg, 265
Mule wagons, 135
Murray, Lance-Corporal James VC, 96
Musket, 20, 25, 281, 284
Musket Wars, 25
Musketry, 24, 231

N
Naauwpoort, 265-266, 268
Naauwpoort Junction, 265
Nascent doctrine, 14, 79, 90, 283
Natal, 41, 45, 52-53, 55, 58, 86, 93, 95, 124, 126, 138, 186, 191, 211-212, 229, 251
Natal Field Force, 37
Natal Front, 42, 48
Natal Mounted Police, 141
National and Historical Monuments Act, 261
National Monument, 97, 100, 106, 110, 116
National Scouts, 73, 79, 240
Nationalism, 39, 88, 237
Native, 13, 20, 22, 35, 39, 52, 69, 77, 80, 89-90, 93, 141, 151, 171-172, 193-194, 234, 282
Native American, 20, 22
Native Scouts, 80
Natural and Historical Monuments, Relics and Antiques Act, 261
Naval Brigade, 38
Ndebele, 36
Nelson, Admiral Lord Horatio, 233
Nelspruit, 48, 118, 199
Nepotism, regimental, 152
Netherlands, 20, 34
New Brunswick, 24-25
New England, 20
New Model Drive, 72-73, 76
New York, 21, 189
New Zealand, 20, 25-26
Newcastle, 53, 191, 266
Night patrols, 202, 204
Night-guard, 126, 154
North Maine, 22-23
North West Frontier, 229
Norvals Pont, 107-108, 212, 231
Noupoort, blockhouse, 117

Nubian Desert, 81

O
Obstacles, 52, 85, 154, 202, 210
Officers' Mess, 16, 236
Olifants Nek, 268
Oliver's Hoek, 269
Omdurman, 235-236
Ontario, 24
Orange Free State, 12, 35-36, 40-42, 45-47, 52-53, 57, 65, 73, 88-89, 125, 140, 151, 212, 215, 225
Orange Grove, blockhouse, 252, 260
Orange River, 42, 46-47, 50, 66, 68, 107, 155, 167, 209, 212, 231, 256
Orange River Station, 231
Order of Battle, 41, 129
Orders, 39, 70, 75, 80, 96, 123, 160, 165-166, 200-202, 210, 236
Other Ranks, 41, 55, 192, 200
Ox-wagon, 51, 125-126, 130, 135, 175, 179-180, 184-186, 188, 217

P
Pakenham, Thomas, 211
Palisades, of stockade 20
Palmerston Forts, 19
Parliament, British, 70, 174, 218, 232, 261, 263
Pastoral care, 206
Patriotic War, Canada, 23
Patrols, 141, 149, 202-204
Pedi, 172
Peninsular War, 13
Petrusburg, 57-59, 265
Pietersburg, 53, 60, 225, 265, 269
Pillbox, 190, 284
Pittsburgh, America, 20
Polfontein, 269
Policy of the blockhead, 211, 222
Politicians, 38-39, 49, 237
Pondicherry mercenary, 93
Port Alfred, 53, 138
Port Elizabeth, 138, 140, 143, 145
Portuguese East Africa, 54
Potchefstroom, 185, 267
Pretoria, 133, 140, 206, 226, 232, 256

Defence of, 59-60, 96-97, 100, 117, 215, 243
Forts, 41, 183
Occupation of, 118, 123
Strategic objective, 42, 45-48, 53-56, 64, 79, 213-215, 229, 231
Town defences, 265
Prieska, blockhouse, 115
Prisoner of war, (POW), 161, 180, 218
P'tit Blockhouse, Canada, 247
Purvis, Major JD, 133

Q
Quaggafontein, 266
Queen Victoria, 35, 49, 98, 197, 230, 234-236
Queen's Blockhouse, 92
Queen's Regiment, 201-202
Queen's South Africa Medal, 145, 250, 253
Queenstown, 138, 266

R
Radio, wireless, 151, 165-169
Railway lines,
    Armoured trains, 63, 79-80
    Attacks on, 55
    Before the war, 52-53, 82
    Blockhouses, defence, 71-72, 86, 102, 110, 117, 123, 131, 140, 161, 166, 180, 205, 241, 243-244
    Blockhouse effectiveness, 211-212, 216-219, 249
    Cable laid for, 166
    Cape Colony, 49, 57, 103, 143
    Concentration camps, 50
    Cost of, 221
    Cycle, railway, 206-207
    De Wet, hunting of, 84, 184-184
    Egypt, 29, 81, 235, 246
    Operations of, 82, 120, 151
    Pretoria to Delagoa Bay, 48, 55, 225
    Strategic importance, 231-232
Railway cycle, 206-207
Railway siding, 63, 80, 84, 216
Ramparts, 93, 96, 113, 142
Rate of degradation, 219
Rations, 73, 136, 169-170, 197-198
Rawlinson, Colonel HS, 78, 180-181, 183

Red Fort, 27
Redoubt, 60, 100, 115, 123, 142, 247, 284
Refugees, 64, 69, 150, 242
Reitz, Denys, 34, 46, 223, 226
Reprisal, 48, 65, 232
Reserves Act 1977, 26
Revetment, 122, 127-128, 135, 273, 284
Rice, Major Spring, 17, 61, 63, 100, 117-118, 120, 123-125, 132, 214, 251-255
Rideau Canal, 24
Riet River, 59
Rifle racks, 153-155, 217, 249, 271-272
River crossings, 61, 56, 118, 120
Riversford, blockhouse, 107-109
Roberts, Lord Frederick,
   Advance on Pretoria, 46, 49
   Anglo-Afghan War, Second, 27, 65, 229, 233, 235
   Blockhouses, & defences, 56, 99, 117, 123, 124, 224, 241-243
   Farm burning, 49-50, 65-66, 232, 240
   Indian Ring, 152
   Key town captures, 58, 64, 79, 123, 215, 231-232
   Kitchener, Lord Horatio, 238
   Majuba, 230
   Rail movement of troops, 82, 216
   Rawlinson, Colonel HS, 180
   Refugee camps, 68, 70
   Relief of siege towns, 44-45, 53
   Out of retirement, 43
Robertson, 265
Rocky Mountains, 22
Rolling machine, 120, 123-124, 185
Rolling stock, 212-213, 216
Rooi Nek, 137
Rosmead, 265-266
Rowntree, confectionary company, 197
Royal Air Force, 189
Royal Arsenal, 81
Royal Artillery, 78, 80, 229
Royal Engineer, 28, 56, 86, 99-100, 102, 105, 112, 118, 123, 126, 137, 180, 220, 236, 248, 252-253, 258, 272
Royal Military Academy, 229, 233, 245, 251
Royal Navy, 166-167
Runner, 165, 200

Russia, 35, 240
Russo-Turkish War, 234

S
Sabotage, 30, 48-50, 55, 82, 213, 215, 224
Saint Andrews Blockhouse, Canada, 24
San Juan Hill, 33
Sanna's Post, 267
Santiago, 30, 157
Sapper, 71, 80, 100, 135, 166, 247, 284165,
Scandinavia Drift, 267
Scheepers, Commandant Gideon, 49
Scorched Earth, 14, 51, 64, 66, 68, 90, 150, 173, 223, 237-238, 256, 285
Scouting, 37, 241
Seaforth Highlanders, 94, 181
Searchlight, 83-86, 186
Sebastopol, 100
Second Anglo-Afghan War, 27, 49
Semaphore, 94, 162, 164
Sergeants' Mess, 200
Sheriff of Nottingham, 113
Shield, protective, 124, 127-128, 273, 285
Siah Sung, 27-28
Sieges, Boer, 41-42, 46-47, 53, 96
   Kimberley, 42
   Ladysmith, 45, 120, 252-253
   Mafeking, 46
   O'Okiep, 145
Siege of Khartoum, 235
Siege of Santiago, 157
Siemens and Halske, 167
Signal Rockets, 126, 161, 203
Signal Tower, 118
Simon's Town, 19, 93, 167, 265
Sir Lowry's Pass Extn, 265
Smatdeel, 265
Smith, Lucian B, 157
Smuts, General Jan, 34, 46, 223, 226
South African Constabulary, 57, 101, 138-140, 216
   Post lines, 265-268, 270
South African Defence Force, 55
South African Heritage Resources Agency, 262
South Wales Borderers, 260
Soutpansberg, 98

Spain, 13, 20, 29, 33
Spanish War, 32
Spanish-American War, 33
Spiders-web wire, 157
Spies, 65, 78, 239
Spring gun, 160, 194
Springfontein, 103, 145, 265
Springs, 60, 178, 266
SS Britain, 239
SS Canada, 243
SS Dunottar Castle, 231
St Omer, 233
Standerton, 71, 133, 191, 212, 222, 266-268
Standerton Line, 267
Standing orders, 96
State Artillery, 41
Station, rail,
    Elandsfontein, 133
    Hanover, 225
    Komatipoort, 54
    Karee, 58
    Krugersdorp, 56
    Orange River, 231
    Reinforcement of, 80, 84. 103
    Riversford, 59
    Val, 124, 267
Steam, 84, 175-178, 187-188, 225
Stellenbosch, 265
Sterkstroom Junction, 266
Stock, 50, 65, 197, 212-213, 216
Stockade, 20, 22, 285
Stormberg Junction, 42, 265-266
Stormberg Junction, blockhouse, 62, 103, 106
Strategy,
    Blockhouse component, 21, 224
    Boer, 213, 215
    British, 16, 34, 52, 90, 145, 148, 237, 257
    Components of British Strategy, 64
    Farm burning, 70
    On-the-fly, 14, 79
    Scorched earth policy, 66
    Sun Tzu, 52
    Trocha, Spain, 29-32
    Initial phase undoing, 43
Strydpoort, 267
Suakin, 28-29, 31, 100, 118, 123, 235, 246-247, 249-250
Sudan, 28, 78-79, 82, 172, 231, 235-236, 246, 248
Sudanese Military Railway, 81
Suez Canal, 246
Supply wagons, 134, 183
Surrender, 47, 85-86, 243, 281
Swazi, 171-172
Swedish, 92
Swellendam Extension, 265
Switzerland, 233-234

T

Tafelkop, 184, 268
Tank
    British Mk IV tank, 189
    German A7Vs tank, 189
Telegraph, 65, 80, 83-84, 117, 120, 151, 163-168, 196, 217, 220-221, 271-272, 284
Telephone, 164-166, 284
Thaba 'Nchu, 267
Tiger's Eye stone, 115
Tinned food, 136, 198
Tomlinson, Richard, 113
Tommy Atkins, 194, 204
Town Guard Forts, 101, 106, 137-138, 143
Train attacks, 48, 53-54, 61, 63-64, 118, 123, 213-214, 232
Transvaal, 12, 35-38, 40, 48, 50, 52-55, 57-58, 60-61, 65, 68, 88-89, 96, 124, 129, 140, 151, 212, 215, 225-226, 239, 244, 256, 258
Transvaal burgher, 241
Transvaal Rangers, 241
Treaty of Vereeniging, 88, 125, 240, 256
Treaty of Waitangi, 25
Trocha, 29-32
Tugela River, 42, 96
Typhoid, 45, 173, 204-205, 226

U

Uitlanders, 40
Uitskud, 66, 285
Uniondale, fort, 106, 143-144, 263
United States, 20, 32-33, 157
Upington, 174
Upper Hutt Blockhouse, New Zealand, 26

V
Vaal River, 46, 112, 268
Val Station, 124, 267
Valsch River Bridge, 268
Van Reenen's Pass, 268
Ventersdorp, 184-185, 268
Vereeniging, 111-112, 125, 212, 265
Vet River Bridge, 267-268
Victoria Cross, 229-230
Victoria Road, 268
Victorian era, 35, 49, 151, 233
Viljoen, General Ben, 55, 84, 219
Villiersdorp, 266-267
Vlakplaats, 266
Volksrust, 269
Vredefort Road Koppie, 116
Vryburg, 133, 258
Vryheid Extension, 267
Vryheid Town, 269

W
Wadi Halfa, 81, 235
Waitara, 26
Wakkerstroom, 267, 269
Waldrift, blockhouse, 111
Wallaceville Blockhouse, 26
Walled City, 285
War Office, 53, 166, 176-177, 245, 270
Warmbaths, 206
Warrenton, blockhouse, 111, 263-264
Washington State, 22
Watchtower, 30, 276
Water cart, 105, 170
Waterloo, 13, 231
Wellington, 62, 245, 248
Wellington, Duke Arthur Wellesley, 233
West Fort, 97
West Rand, 60
West Redoubt, 247
White Man's War, 13, 88, 171
White, Sir George, 253
Whitworth, 6-pounder, 41
Wilge River, 266
Williston, 71, 269
Winburg, 265
Windmill Hill, 117
Winston Churchill, 32, 100, 200, 236

Witkop, blockhouse, 112-113, 197
Witkopies Post, 137, 157
Witwatersrand, 39, 58, 127
Wolseley, blockhouse, 248
Wolseley, General Sir Garnet, 152, 229-230, 232-234
Wolverhampton, 127
Wolwehoek, 266
Women's Committee, 173
Wonderfontein, 268
Wood, Adjutant-General Evelyn, 167
Wood, Major-General Sir Elliot, 28-29, 57, 60-63, 93, 100, 102-105, 108, 110, 115, 118, 123-124, 128-129, 131, 183, 221, 224-225, 244-247, 250
Woolls-Sampson, Colonel Sir Aubrey, 78
Worcester, 265

X
Xhosa, 36, 93-94
Xhosa Wars, 13

Y
Yakima Indian War, 22
York & Lancaster Regiment, 136

Z
Zand Bridge, 267
Zanzibar, 235
ZAR, 12, 40, 48
Zulu, 13, 16, 25, 36, 68, 95-96, 135, 230, 282
Zuurfontein, blockhouse, 107, 260
Zephyr Canoe, 245

www.ingramcontent.com/pod-product-compliance
Lightning Source LLC
Chambersburg PA
CBHW061804290426
44109CB00031B/2931